KARGER 125 YEARS

Harold M. Schmeck, Jr.
Mechthild Hempe

KARGER 125 YEARS
Medical and Scientific Publishers

Contents

Connecting the World of Biomedical Science

125 years – what an impressively long period of time. It is indeed a great pleasure for me to present this Festschrift to you and I trust that it will give you an interesting and accurate picture of how our company has developed over the years. It was clear to me from the outset that, given my deep love of history, I was very much in favor of producing a sequel to the volume Harold M. Schmeck had put together for our Centenary Celebrations.

The Festschrift *Karger – Turning Medical Progress into Print (1890–1990)* is structured around four generations that show how Karger Publishers and my ancestors managed to continuously maintain the company even in times of extreme political threat and economic crisis – and yet somehow they managed not only to survive, but to prosper despite the many challenges that faced them. With its insights into the history of the period, of our publishing company and the Karger family as well as the history of medicine this book has itself become a historical document to which nothing needs to be added.

My gut feeling had initially led me to expect that the last 25 years would not stand comparison with the 100 years that had preceded them, but I was wrong. From a historiographical perspective the years since 1990 have been anything but tranquil. The 9/11 attacks come to mind, as do the end of

Apartheid, the digital revolution, Dolly 'the world's most famous sheep' and the introduction of the Euro currency, to name just a few of the events that stand out in recent memory. Technological advances have continued to develop at a mind-blowing pace, overshadowing just about everything that was viewed as a major achievement in bygone years.

It therefore quickly became clear that the best way forward for our project would be to retain the original text from 1990 by Harold M. Schmeck and update it by adding the history of the last 25 years. The focus of the new part was, however, to be somewhat different to the approach that had been adopted for chapters 1 to 6: the sequel mainly looks at the challenges being faced by the entire publishing world and investigates how knowledge is communicated at the end of the 20th and at the beginning of the 21st century.

We then asked the 'Geschichtsbüro' in Cologne to compile the new section for our 125th Anniversary Festschrift. Dr. Mechthild Hempe conducted interviews with my father Thomas Karger and also with Thomas Nold, Director of the Karger Publication Program, Moritz Thommen, Director of Sales and Distribution, Mike Dürst, Director of Production and Publishing Technology, and Annemarie Gohl, Director of Human Resources, as well as with me. Based on the information she was able to glean from these discussions and from other sources such as our staff newsletter *Allerlei*, she succeeded in covering the most significant developments at Karger Publishers over the last 25 years. Her account has been organized into chapters 7 to 9.

I would like to express my gratitude to Dr. Hempe for her input. I am similarly indebted to all members of staff who placed their knowledge and expertise at our disposal. A special thanks goes to Thomas Nold, who coordinated the entire project. Gerri Schmid, senior graphic designer at Karger, developed the new layout and used his many years of experience and skills to ensure that the original publication covering the first 100 years smoothly connected up with the new text dealing with the last 25 years.

I would like to draw your attention to the introduction my father wrote for the 1990 Festschrift. Many of the phrases he used remain as valid today as they were then. I know that I can speak on behalf of everyone working in all of the different divisions and departments in this company when I say that we are profoundly grateful for his leadership and support as we continue to uphold this successful company in the second century. All of us would also like to congratulate him on his 85th birthday, which happens to coincide with the company's 125th anniversary. We wish him continued good health and much happiness.

And now I would like to wish you a lot of pleasure in reading – irrespective of whether you start with the last 25 years or at the very beginning of the complete history of Karger Publishers.

Gabriella Karger

Harold M. Schmeck, Jr.

Karger – Turning Medical Progress into Print

A Mirror of a Century of Medical and Scientific Publishing

Introduction to Karger – Turning Medical Progress into Print

It is with great pride that I present this festschrift on the occasion of the centenary of the foundation of S. Karger.

Mr. Harold M. Schmeck, Jr., who came highly recommended to me as a longstanding, respected retired science correspondent for the New York Times, noted for both objectivity and clarity, accepted to write *Karger – Turning Medical Progress into Print.* I would like to express to him my great appreciation and my sincere and warm thanks for his research and attentive efforts to give an account of our company's evolution under the guidance of three generations of the Karger family – and recently the fourth, and to place the events into historical perspectives.

The history of an enterprise, related by a stranger, thus from an external view, will obviously give a unique, while at the same time a personal and subjective picture, all the more so as Mr. Schmeck was given complete journalistic freedom. We believed that this unbiased outside view would generate welcome reflections on our past.

Clearly, an author living in the United States would treat such a subject from an American point of view: the individuals quoted thus mainly belong to his own environment, leaving the Europeans somewhat under represented. But we were interested in a lively story, based on conversations with the protagonists, rather than on a dry compilation of documents.

My sincere thanks go to Mrs. Greti Schwärzel as well as to Mrs. Mary-Kay Kindhauser, who assumed with great talent the significant task of introducing Mr. Schmeck to our organization, to Mrs. Marie-Louise de Stachelski, who reliably carried out the copy-editing, to Mr. Erich Gschwind for his outstanding achievement as a designer, to the production team for their valuable work under the supervision of Mr. Alexander Geraets, as well as to Mr. Roberto Zoppi for the demanding task of coordinating the production and publication of this commemorative volume.

Our company owes its existence and continuance primarily to the collaboration of dedicated editors, authors and advisors, and to the support of its staff, who have maintained the high quality standards which have always been the hallmark of S. Karger. I extend my deepest and personal appreciation to all of them.

I also wish to thank my wife, Yvette, who, through her steady and faithful support during the thirty-two years of our life together, helped me to carry my responsibilities and has thus, in her own way, played an important part in the success of the company.

It has been – and is – an essential philosophy of S. Karger to contribute to scholarship and education in a healthy, straightforward business manner, focused on highly respected publications for a professional audience. All those connected with this house wish to persist along these lines stay modest, pay attention to one's work and avoid the pitfalls of growth and expansion for self-serving aggrandizement.

This history of S. Karger shows that in spite of political and economic adversities, and their impact on science, the company's publishing policy has been a good and rewarding one. To conclude, I hope I will have the privilege to go on 'turning medical progress into print' and to lead the fourth Karger generation and our company into the next century.

1

The Birth of an Idea

1890

The final years of the 19th century were a time of ferment and amazing progress in medicine. In 1890, Europe was at peace. The great empires seemed to be learning how to get along with one another and the world was full of hope and promise. Nowhere was this more true than in the art and science of medicine.

Like the professional climate a century later, practice was being revolutionized by a drumfire of startling advances that had occurred in the previous decades to change the entire world of science.

Ignaz Semmelweis had died his lonely death only 25 years earlier, after demonstrating that doctors' cleanliness could save women from death by puerperal sepsis.

Practical use of the theory that germs caused most of mankind's deadly scourges was scarcely 25 years old, but was beginning to make its influence felt powerfully in the

Ignaz Semmelweis
(1818–1865)

Theodor Schwann
(1810–1862)

Robert Koch
(1813–1910)

treatment of disease. Even the revolutionary concept that all plants and animals were 'aggregates of cells which are arranged according to definite laws' had been stated less than a half century earlier by the great German zoologist Theodor Schwann. Robert Koch, a pioneer of modern bacteriology, had only in 1876 proved that a bacillus was the sole cause of anthrax and, several years later, had isolated the vibrio that causes cholera. Shortly before 1890, he had been given control of a new and powerfully equipped institute in Berlin.

The faculties of the medical institutions in that city were filled with famous names. Germany, and particularly Berlin, was at the heart of the new era of medical progress. Doctors everywhere needed to keep up to date with all the marvels medical scientists were discovering.

This was the climate in which Samuel Karger, a young bookseller who had recently come to Berlin from Leipzig, decided to go into business himself as a publisher. He was only 27 years old, but already a vigorous and dynamic man who had a strong business sense and a remarkable talent for meeting people and gaining their confidence.

Samuel Karger with his son Heinz.

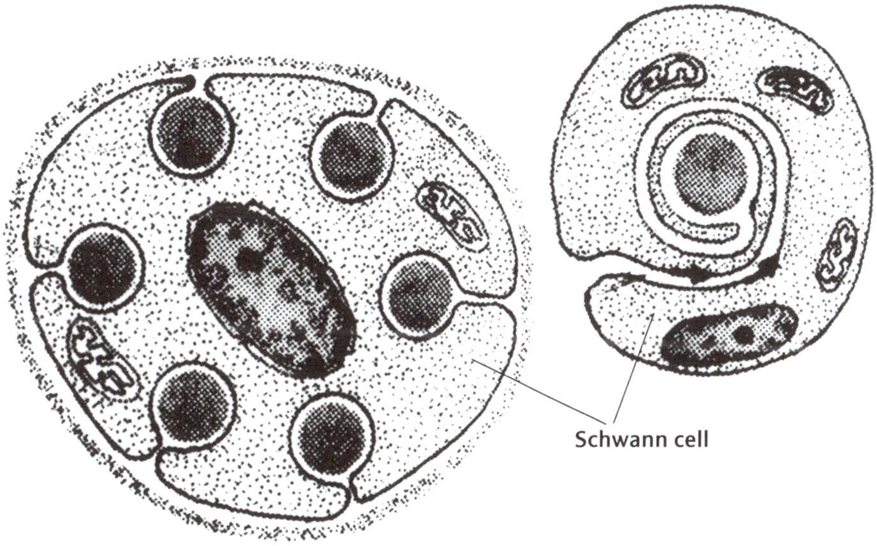

C.F. PRELL'S

GUSTAV FOCK

Erteilung litterarischer Auskünfte und Ratschläge.

Stuhr'sche Buch= & Kunsthandlung

Most of the records of his early years have been lost. Of his ancestors it is known only that they were servants at Castle Karge in Poland, an estate from which their name was derived. No one today knows just why Samuel Karger chose bookselling as an occupation. It is known that he began an apprenticeship in bookselling in the bookstore Ebbeke, then the largest in the city of Posen (now Poznan, Poland). In his later years, he liked to recall an anecdote from that early time. It was an exchange he had with a veteran bookseller before Christmas, one year when the young apprentice was unpacking a shipment of new books.

He was particularly taken with one volume and showed it to his supervisor, remarking that it was indeed 'an exceptionally beautiful book'. The supervisor quickly reminded him of the realities of life in his chosen occupation: 'A book is beautiful, my dear Karger, when it is sold', said the older man.

Evidently, the young apprentice took that admonition very much to heart. The certificate given him by the bookstore at the end of his apprenticeship at Ebbeke contained this remark: 'When someone came into the bookstore asking for a cheap paperback, Karger ended up selling him an encyclopedia.'

After the apprenticeship, the young Karger worked in several other cities, including Lucerne, where he was a first assistant at the bookstore Prell. In Lucerne he grew a beard, he later told his son Heinz, so that he would look older and make a properly dignified and imposing impression on his clients. Later he moved to Leipzig where he was again a first assistant at another major bookstore: the Gustav Fock Company. After Leipzig he came to Berlin, to the bookstore Stuhr, one of the most prestigious in that city.

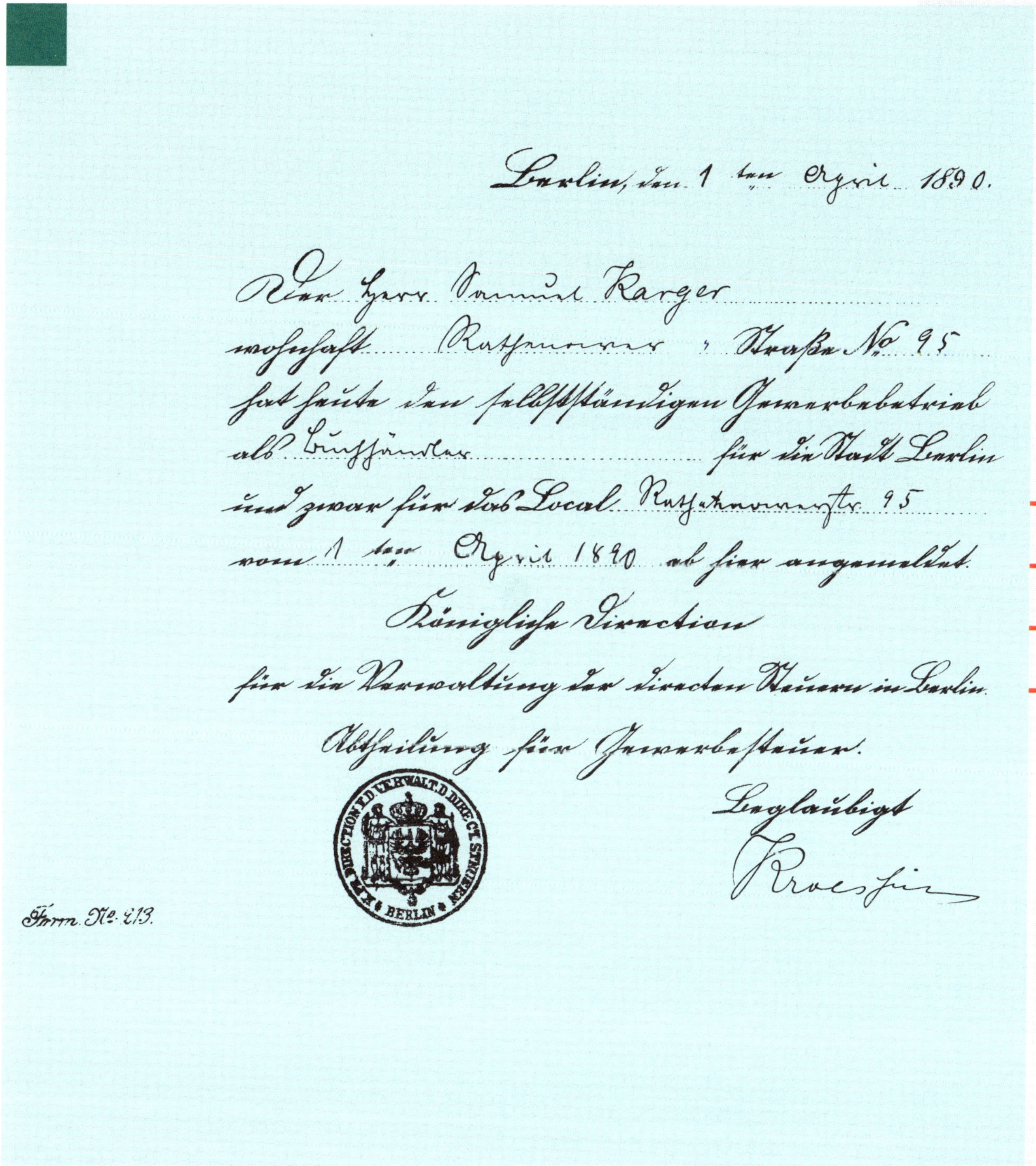

Berlin, den 1ten April 1890.

Der Herr Samuel Karger
wohnhaft Rathenower Straße No 95
hat heute den selbstständigen Gewerbebetrieb
als Buchhändler für die Stadt Berlin
und zwar für das Local Rathenowerstr. 95
vom 1ten April 1890 ab hier angemeldet.

Königliche Direction
für die Verwaltung der directen Steuern in Berlin
Abtheilung für Gewerbesteuer.

Beglaubigt

Form. No. 213.

Document in connection with the establishment of the publishing house. 1st of April 1890, Rathenowerstrasse 95, Berlin

‘I intended to make a collection of compendia, the various sections of which would cover the whole field of medicine.’

Samuel Karger (1863–1935)

Charité Hospital, Berlin, 1910

It was in Berlin that he gave up selling books for other publishers, and launched his own company. His plan was to specialize in medicine and science.

For an ambitious young man seeking to make his mark in publishing, medicine was an inspired choice. Berlin was a major center of medical and scientific progress in that time, and his brother Max was already studying medicine there.

Samuel Karger's idea, apparently highly unusual in that era, was to publish a series of compact, but definitive, manuals which would survey and digest the state of the art of important fields of medicine of that time. The intended audience included not only students, but practising physicians as well. The books were not to be simply 'crutches' to help medical students at examination time, but thorough, practical and up-to-date reviews of the major fields: 'I intended to make a collection of compendia, the various sections of which would cover the whole field of medicine', Samuel Karger wrote many years afterwards. 'Not just mnemonic devices for the exams, but manuals which would deal briefly and precisely with each theme in a didactically perfect manner.' He envisaged manuals useful for students and concise, practical reference works – 'vademecum' was the term he chose – that would be useful to practising physicians.

The first years must have been hard. The young would-be publisher set up his business in the same rooms where his brother studied. It was a small office in a building that was bounded on one side by Berlin's Charité Hospital and on the other by the city's Royal Gynecology Clinic. Appropriate to that setting,

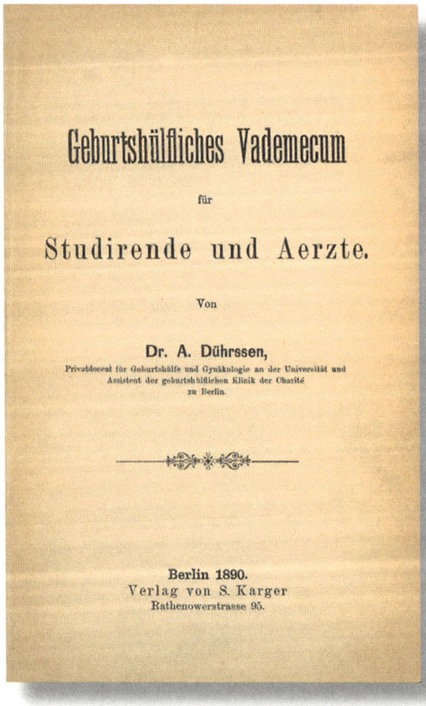

Geburtshülfliches Vademecum

für

Studirende und Aerzte.

Von

Dr. A. Dührssen,

Privatdocent für Geburtshülfe und Gynäkologie an der Universität und
Assistent der geburtshülflichen Klinik der Charité
zu Berlin.

Berlin 1890.
Verlag von S. Karger
Rathenowerstrasse 95.

the first definitive manual under the Karger imprint was on obstetrics. It was titled *Geburtshülfliches Vademecum* [1], and the first printing, in 1890, was sold out in less than six months. The book went through many editions. With it, the young company was off to a solid and prosperous start.

Apparently, Samuel Karger was never able to forget that he had to leave his own formal education incomplete and go to work to support his brother and sister. But that seemed to strengthen his determination to succeed and to create something worthwhile. One of the satisfactions of publishing, particularly in a field like medicine, is the assurance that a quality product is more than simply something to be used or consumed. Such a product serves an important purpose. It is potentially useful to all of mankind: capable of saving lives, preventing or curing disease, relieving suffering and adding to the sum of significant human knowledge. Indeed, this concept has guided four generations of the Karger family and has established the standard by which their books and journals are published. With the devotion to high scientific quality came a high regard for excellence in typography and all the other aspects of quality book design. Samuel Karger had expressed a real and deep-seated appreciation of fine books in that early exchange with the bookstore manager.

Probably the close association with his brother and perhaps other medical students helped channel the publisher's thoughts into the most promising areas of medical publishing. In any case, his idea was welcomed by young medical scientists who were already involved in research and teaching. From among them he was able to recruit expert advisors and highly competent authors.

The plan quickly became a remarkable success, partly no doubt because of a wise choice of authors. The first book, the obstetrics manual, was by A. Dührssen [1], a specialist in obstetrics and gynecology. It was published in 1890 together with a companion volume, *Gynäkologisches Vademecum* [2], by the same author. That too was received avidly by the medical community. Surviving records of the company show that the first book went into a fifteenth edition by 1909, the second in a sixteenth edition by 1922. Originally published in German, the two books were later translated into English, French, Italian, Polish, Russian, Greek and Turkish.

A MANUAL OF

GYNÆCOLOGICAL PRACTICE

Vade-Mecum de Gynécologie

ΓΥΝΑΙΚΕΙΟΛΟΓΙΚΟΝ ΕΓΧΕΙΡΙΔΙΟΝ

Gynäkologisches Vademecum

KRÓTKI PODRĘCZNIK

chorób kobiecych

During the young company's first ten years, a substantial number of other concise manuals were also published in a broad range of medical specialities, including P. Silex's *Compendium der Augenheilkunde* [3], published in 1891, *Compendium der allgemeinen* (1892) *und speziellen* (1893) *Chirurgie*, by G. De Ruyter and E. Kirchhoff [4]. In 1898, Karger published the *Compendium der Physiologie des Menschen*, by P. Schultz [5] while R. Kayser and W. Klestadt's *Anleitung zur Diagnose und Therapie der Kehlkopf-, Nasen- und Ohrenkrankheiten* [6] appeared in 1901. All of these early books went into several editions and were translated into many languages.

Soon after the publishing house of S. Karger produced its first books and journals, it became a center for the medical sciences in Germany and abroad: 'Specialists at famous university clinics in Berlin, Vienna, Leyden, Heidelberg, Paris, published their books with Karger or cooperated on the various periodicals he founded', a colleague from another company noted many years later. The publications were used in universities all over Europe and even beyond the continent.

The time was obviously ripe for an imaginative new venture in medical publishing. In the company catalogue published on the 25th anniversary of the founding, Samuel Karger noted that medicine had been in a period of rapid progress at the end of the 19th century: 'Bacteriology, antisepsis, asepsis, chemotherapy, and more recently, radiology, radiotherapy and phototherapy aroused an extraordinary amount of reports which needed new publishing media', he wrote. 'New journals were created and readily marketed as all the foreign countries endeavored to take advantage of the advances in German science.'

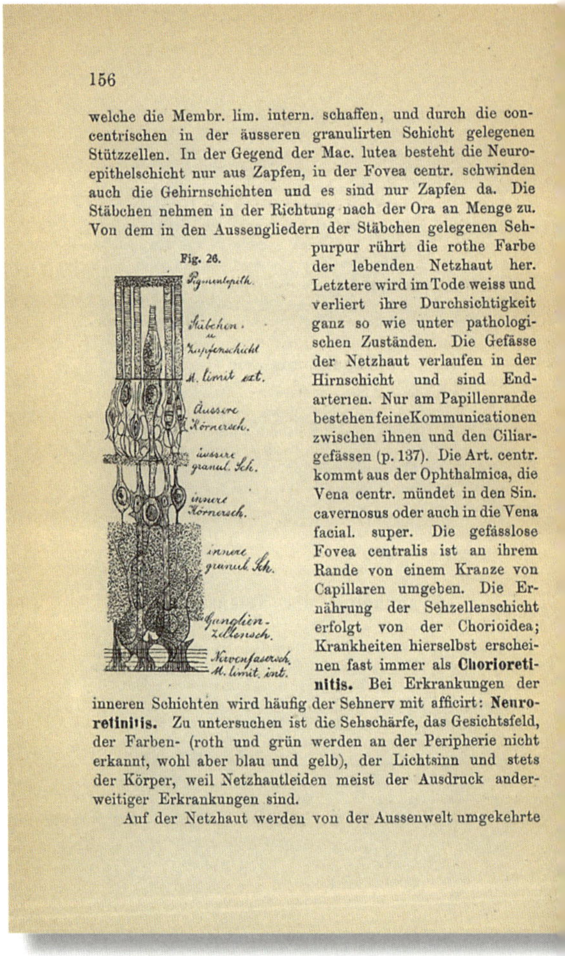

156

welche die Membr. lim. intern. schaffen, und durch die concentrischen in der äusseren granulirten Schicht gelegenen Stützzellen. In der Gegend der Mac. lutea besteht die Neuroepithelschicht nur aus Zapfen, in der Fovea centr. schwinden auch die Gehirnschichten und es sind nur Zapfen da. Die Stäbchen nehmen in der Richtung nach der Ora an Menge zu. Von dem in den Aussengliedern der Stäbchen gelegenen Sehpurpur rührt die rothe Farbe der lebenden Netzhaut her. Letztere wird im Tode weiss und verliert ihre Durchsichtigkeit ganz so wie unter pathologischen Zuständen. Die Gefässe der Netzhaut verlaufen in der Hirnschicht und sind Endarterien. Nur am Papillenrande bestehen feineKommunicationen zwischen ihnen und den Ciliargefässen (p. 137). Die Art. centr. kommt aus der Ophthalmica, die Vena centr. mündet in den Sin. cavernosus oder auch in die Vena facial. super. Die gefässlose Fovea centralis ist an ihrem Rande von einem Kranze von Capillaren umgeben. Die Ernährung der Sehzellenschicht erfolgt von der Chorioidea; Krankheiten hierselbst erscheinen fast immer als **Chorioretinitis.** Bei Erkrankungen der inneren Schichten wird häufig der Sehnerv mit afficirt: **Neuroretinitis.** Zu untersuchen ist die Sehschärfe, das Gesichtsfeld, der Farben- (roth und grün werden an der Peripherie nicht erkannt, wohl aber blau und gelb), der Lichtsinn und stets der Körper, weil Netzhautleiden meist der Ausdruck anderweitiger Erkrankungen sind.

Auf der Netzhaut werden von der Aussenwelt umgekehrte

Fig. 26.

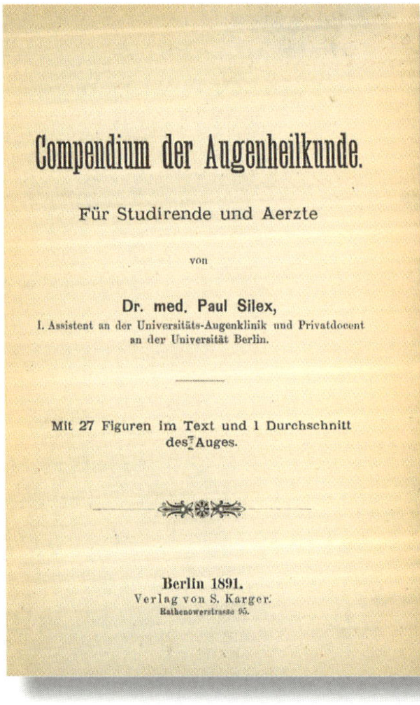

Compendium der Augenheilkunde.

Für Studirende und Aerzte

von

Dr. med. Paul Silex,

I. Assistent an der Universitäts-Augenklinik und Privatdocent an der Universität Berlin.

Mit 27 Figuren im Text und 1 Durchschnitt des Auges.

Berlin 1891.

Verlag von S. Karger.

Rathenowerstrasse 95.

157

reelle Bilder entworfen. Es findet durch den Einfluss des Lichtäthers dabei eine chemische Umsetzung des Sehpurpurs und wahrscheinlich anderer Sehsubstanzen statt, durch welchen Vorgang die zum Gehirn führenden Nervenfasern erregt werden, wodurch alsdann eine Empfindung und Wahrnehmung ermöglicht ist.

Die Krankheiten zerfallen in solche **mit organischen** und solche **ohne organische Veränderungen.** In die erste Gruppe gehören:

1. Die **Netzhautablösung** s. Sublatio retinae. Diagnose nur ophthalmoskopisch möglich, am leichtesten mit der blossen Durchleuchtung, da durch die Vortreibung meist hypermetropische Refraction gebildet wird. An den nicht abgelösten Partien haben wir den bekannten rothen Schein, an den abgelösten eine bläulich-graue, gefaltete, oft flottirende und mit den dichotomisch getheilten Netzhautgefässen besetzte Membran und bei flacher Ablösung nur die schwarzbraun aussehenden und eine Niveaudifferenz darbietenden Netzhautgefässe. Bei einer seichten Ablösung in der Mac. lut. tritt die Fovea als ein rother Fleck hervor. Die Bewegung fehlt, wenn ein Tumor der Netzhaut fest anliegt. Die Verfärbung ist bedingt durch die Fältelung und durch die Beschaffenheit der subretinalen Flüssigkeit, die serös, blutig und eitrig sein kann, Lymphzellen, rothe Blutkörperchen, Pigmente, Fett, Cholestearin und Fibringerinnsel enthält. Die Abhebungen beginnen in der Regel in der oberen Bulbushälfte, von wo aus sich die Flüssigkeit unter Anlegung der Membran bisw. auch unter Rückkehr der Function nach unten senkt. In frischen, schnell aufgetretenen Fällen wird fast immer ein Netzhautriss, kenntlich an seinen scharf umgerollten Rändern und an dem Sichtbarsein der Chorioidea daselbst, gefunden, der nach Anlegung der Membran wieder verschwinden kann. Bei längerer Dauer treten häufig in der abgelösten Partie Bindegewebswucherungen, Pigmentirungen, Entfärbungen des Pigmentes und Blutungen auf. Die meisten Ablösungen sind progressiv, so dass schliesslich totale Ablatio vorliegt und die Ret. nur noch an der Papille und der Ora mit ihrer Unterlage in Verbindung steht. **Complicationen** sind: Glaskörper- und Linsentrübungen, Iridochorioiditis, Ver-

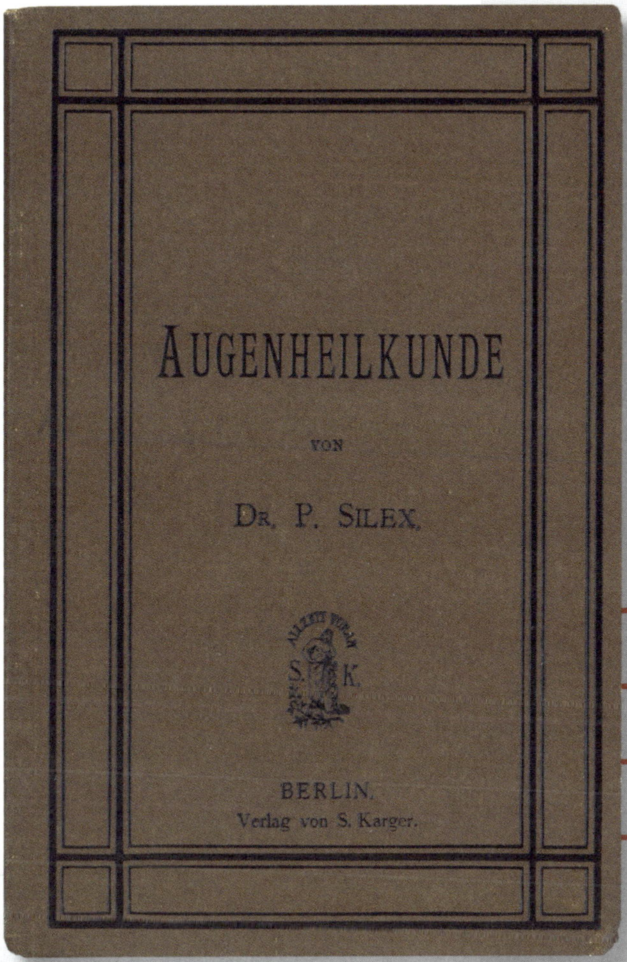

g. 13. (Nach Jäger und Fu

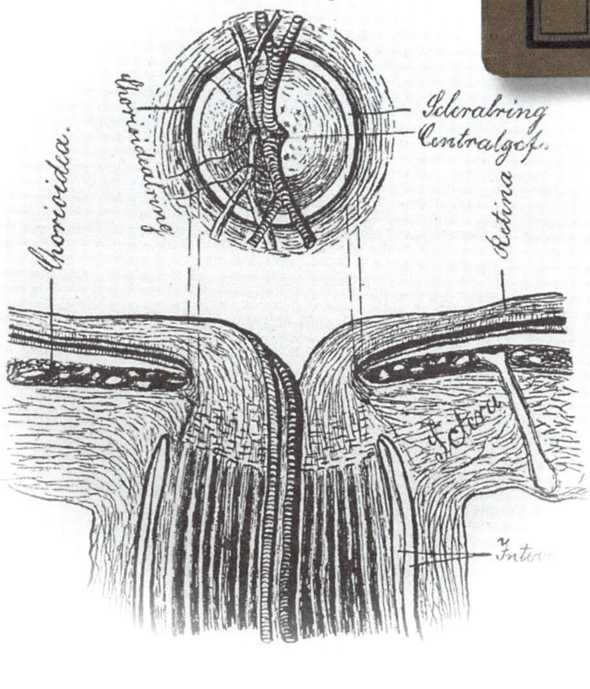

Dermatologische Zeitschrift
Dermatologische Zeitschrift
Dermatologische Zeitschrift
Dermatologische Zeitschrift
Dermatologica
Dermatologica
Dermatologica
Dermatologica
Dermatologica
Dermatology

The young publisher was among those who most quickly saw a market for scientific journals as well as textbooks. He foresaw that medical research must become increasingly specialized as new discoveries opened more and more scientific horizons. He concluded that medical publishers could best serve this expanding world of medical science by producing new journals designed to cover the emerging speciality fields. The publishing house has pursued that strategy successfully for 100 years. His first journal was in dermatology, published in 1893 as *Dermatologische Zeitschrift* [7]. The journal is still published by Karger under the title *Dermatologica* (since 1992 *Dermatology*). Five other journals were added to the publisher's list by the turn of the century, including periodicals devoted to research in gynecology and obstetrics [8], digestive diseases [9], psychiatry and neurology [10], ophthalmology [11], and pediatrics [12].

Twenty-five years after the publishing house was founded, nine journals were being published under the Karger imprint, testimony both to the vigor and initiative of Samuel Karger, and to the rapid expansion of scientific knowledge that marked the whole era. Altogether, the company published about 50 titles that year: 'I was lucky to find men who not only welcomed my suggestions, but who also realized them in such a manner that, especially the periodical projects, developed into the world's most reputed medical works', he wrote.

'I was lucky to find men who not only welcomed my suggestions, but who also realized them in such a manner that, especially the periodical projects, developed into the world's most reputed medical works.'

Publisher's emblem in use from 1892 to 1937

Among the books, one of the most successful over the years was a monumental volume on diseases of the nervous system: *Lehrbuch der Nervenkrankheiten* by H. Oppenheim [13]. Seven editions were published and the work was translated into English, Russian, Italian and Spanish.

In its time it was considered the 'neurologists' Bible'. Samuel Karger often mentioned very proudly that once a whole German edition of this work was sold in Japan.

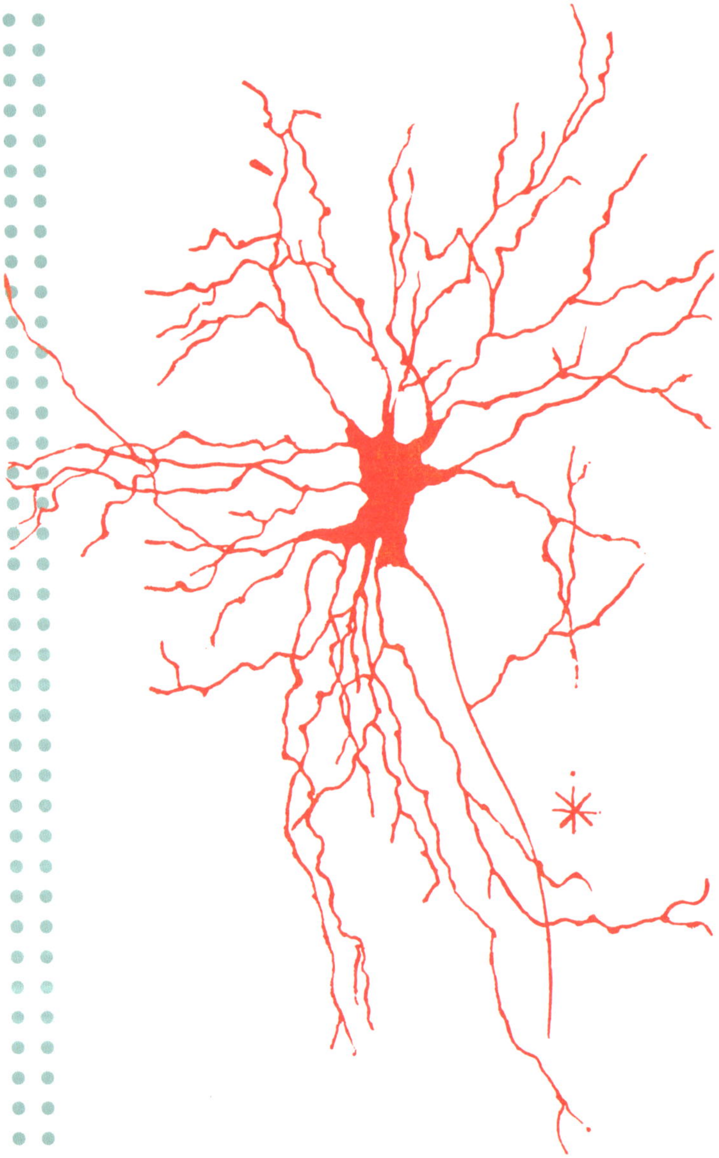

Samuel Karger's personal style was a major factor in the continued expansion of the publisher's list of both books and journals. He maintained an extensive dialogue with his friends in the medical profession and these interchanges kept him up to date on what the market needed. With the objectives in view, the publisher's expert sense of excellence in book design and production completed the equation. His insistence on high standards earned him a place in the exclusive ranks of Germany's top-quality publishers. The medical works, both in process and in print, must have been a focus for some lively discussions among experts who came to the modest but crowded office in the shadow of

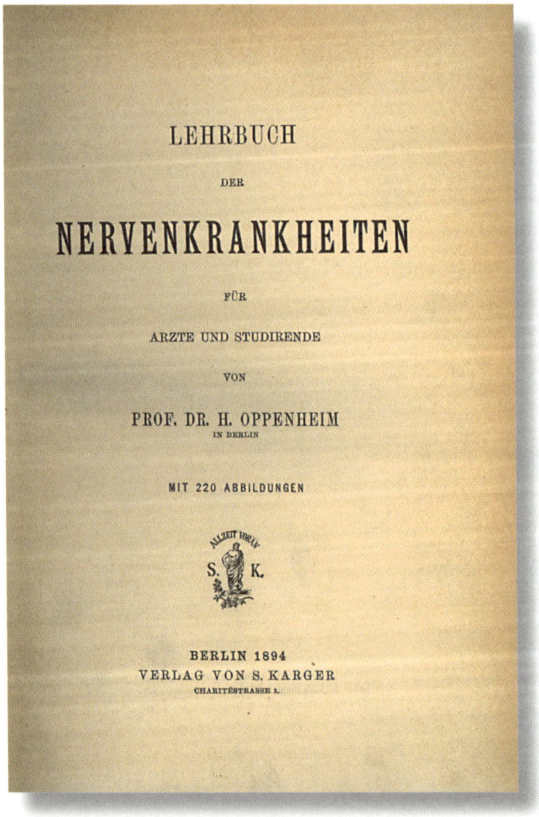

LEHRBUCH

DER

NERVENKRANKHEITEN

FÜR

ÄRZTE UND STUDIRENDE

VON

PROF. DR. H. OPPENHEIM

IN BERLIN

MIT 220 ABBILDUNGEN

ALLZEIT VORAN

S. K.

BERLIN 1894

VERLAG VON S. KARGER

CHARITÉSTRASSE 3.

Samuel Karger in his late sixties

Berlin's major medical centers. It was hardly surprising that the publishing house became a center for informal discussion of emerging trends in medicine and the latest discoveries in science. The publisher began to receive a steady flow of manuscripts, not only from local specialists, but from the medical capitals of Europe.

During the early years of the 20th century, as today, the world of medical science considered itself apart from, and no doubt above, concerns of nationalism and partisan politics. Science was international, in the service of mankind, regardless of the bickering among nations. The introduction to Samuel Karger's 25th anniversary catalogue, published in 1915, made no mention of the devastating war that had begun the previous August.

Science continued even while the nations were at war and, at that time, no one could have predicted the scope of the cataclysm that was just then beginning. Europe had survived many wars. Would one more conflict between the Germans and the French be much different? That insistence on the overriding importance of science actually continued through most of the war. Even in the grim year 1917, with Germany all but defeated, S. Karger began the publication of a new series on psychiatry [14].

Berlin before World War I

But, of course, that war was indeed different. It swallowed nations and warped the course of history. Much as science sought to keep apart from the worlds of politics and economics, this was no longer possible. Samuel Karger had joined the ranks of scientific publishing at a time of great political change as well as scientific progress. The German Empire, dominated and personified by Prussia, had really come into existence in 1871 when Samuel Karger was only eight years old. What must have seemed to him the permanent order of things when he set up his publishing house in 1890 was really the fast-fading end of an era. The new world that was about to emerge would have profound effects on the company and on its stewardship by Samuel Karger's children, grandchildren and great grandchildren.

In 1890, just as the publishing house was binding its first two books, the young emperor William II had dismissed his chief of government, Otto von Bismarck, and started Europe down a path that led to two world wars and a destruction of life and culture that would have been unimaginable to the citizens of Europe during the last years of the 19th century.

The future upheaval must have been impossible to imagine in 1915, with the publishing house making steady progress and all of its journals still successful. The only excep-

Diploma of prize awarded at the International Exhibition in Brussels, 1910

Diploma of prize awarded at the World Printing and Graphics Exhibition in Leipzig, 1914

tion to the pattern of success appeared to have been the journal *Archiv für experimentelle and klinische Phonetik* [15], published in a single volume in 1913 and then discontinued. Members of the family today speculate that this was simply a journal before its time. It was revived in 1957 under the title *Phonetica* [16] and will go into its 47th volume in 1990.

In the 25th-year catalogue, published in April of 1915, Samuel Karger noted with satisfaction that the company was awarded prizes at several exhibitions. These included the Patient Care Exhibition in Berlin, the International Exhibition in Brussels, the Hygiene Exhibition in Dresden and the World Printing and Graphics Exhibition in Leipzig in 1914.

Publisher's Complete Catalogue 1915

'May this friendship stimulate further work to the service of medical science.'

Three quarters of a century later, the very mention of the years 1914 and 1915 evokes the spectre of devastation and tragedy, but apparently when the catalogue was written, the war was not advanced far enough to be seen as the global disaster that it really was. To a medical publisher, the outlook still seemed bright.

'This catalogue, which I publish with great satisfaction, mirrors 25 years of independent activity', Samuel Karger wrote. 'I owe this primarily to the authors of my company, who have placed their confidence in me and thus created the basis for this achievement. Most of them have remained faithful friends, some of them since the foundation of my company. May this friendship stimulate further work to the service of medical science.'

But, even as he wrote the words, the framework of the old Germany and its scientific preeminence were beginning to crumble under the hammer blows of the world's most destructive war. From 1915 onward, the war dragged on, spread, intensified and gradually

Heinz Karger during World War I (1895–1959)

Gerda Karger-Hurwitz (1896–1961)

began to strangle all aspects of life in Germany. Furthermore, Samuel Karger's first son Heinz, who had been born in 1895, was called up to serve as an infantry officer on the Western Front and there was every reason to fear that there might be no one left to carry on either the family or the publishing house. Little remains known of Samuel Karger's experiences in trying to maintain the company during the war or the privations he and his family must have suffered in common with the rest of Germany. Most of the records were lost during the 1930s. It is known that he worried about his son and, with justification. Heinz Karger was wounded in the bitter fighting in Alsace, but he lived through the war, and the publishing house survived the devastating periods of inflation that followed. That success itself was extraordinary, considering the havoc produced in all aspects of German life by the defeat in 1918.

After the war, Heinz Karger completed his studies as an economist and then joined his father in the business; by 1924, the company was returning to normalcy and success. The publisher continued to be successful throughout the turbulent decade of the 1920s. The house of Karger withstood the effects of the financial crash of 1929 and, in 1930, after 40 years of publishing, Samuel Karger could list more than 850 titles that had been produced under his direction. Another joy to the family was the appearance of a grandson. Heinz and Gerda Karger's son, Thomas, was born in Berlin in 1930. At about that time, Heinz Karger began to co-manage the publishing house with his father and helped keep the company afloat through the rigors and economic chaos of the Great Depression.

When Heinz Karger began to assume major responsibility for the company, his father was in his late sixties. Germany was slowly struggling out of the Great Depression. But something new was beginning, against which even the best management skills and devotion to the highest principles of publishing were powerless. In 1933, Adolf Hitler became chancellor. The Austrian with the moustache had made a lot of ranting speeches and had a highly organized corps of thugs at his command, but many intelligent people thought his posturing might he largely propaganda and that, in any case, Hitler could be 'managed'. Like many others, Samuel Karger simply could not believe the worst of what ap-

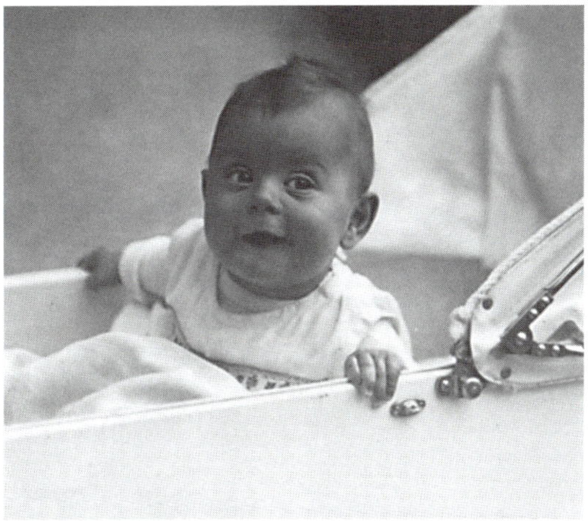

Thomas Karger, 1930

peared to be developing in his country. His son had served in the Army and had been wounded fighting for Germany. Surely that would protect the younger Karger from the malice of the Nazis.

Käthe and Samuel Karger's tombstone at Berlin-Weissensee

Heinz Karger, himself, harbored no such illusions. He had suffered through one session in the Gestapo office in Berlin listening to the insane logic of the new bureaucrats. S. Karger was forced to remove all Jewish scientists from its editorial boards. The young publisher watched as the Nazi officials went through all of the journals, title page by title page, methodically crossing out the names of all 'non-Aryan' editors.

As he told a friend late in his life, Heinz Karger suspected that this attack on the editorial boards was only an early gust of wind in a storm that was sure to grow more violent. He feared that the next blast would take not just the editors, but the owners of the publishing house as well. He concluded that his company – and his family – would have to find a home outside Germany. His father still refused to believe this, which made it virtually impossible for the younger man to plan the escape.

On July 1, 1935, Samuel Karger spent his usual full morning at the publishing house and left in the afternoon to visit a friend and enjoy a game of skat. He particularly liked to visit in the peace of the friend's lovely garden. That afternoon, the family got an urgent telephone call. Samuel Karger had suddenly become very ill. In fact, he died that day at the age of 72, passing quickly and still mercifully unaware of what was about to happen to his country, his family and many of his friends. In a tribute that took real courage in Berlin, in 1935, many of the company's collaborating scientists and editors attended Samuel Karger's funeral at the Jewish cemetery in Berlin-Weissensee.

The next day, Nazi officials ordered Heinz Karger to hand the entire company over to 'Aryans', whose 'racial purity' somehow entitled them to the free gift of a successful publishing house that had taken more than 40 years of hard work to create.

Heinz Karger recalled being happy that his father never lived to see that day. But now the whole burden of the situation had fallen on his shoulders and he had no one with whom to talk about it. In desperation, he began to search for ways out. He tried to persuade the Nazis to make an exception in the case of his company because of the nature and importance of the works S. Karger was publishing. To the Nazis, medical science was not important. He travelled to Zürich, Switzerland, in the hope of moving the company there, but was not able to establish adequate connections. Apparently, he did not know the right people. He made similar trips of desperation to the Netherlands and Czechoslovakia. Those ventures were fruitless too. The disappointments were bitter and frightening. He had no way of knowing how fortunate his failures were. It was impossible to forsee that within only four years, Hitler's evil grasp would reach out to smother Czechoslovakia and that in a scant five years – May 1940 – the Nazis would invade the Netherlands.

Heinz Karger's son, Thomas, was only a young child, but he can still remember the terror of those years.

'My father had terrible worries about his business and every day, very bad news. I recall this very well', Thomas Karger said more than fifty years later. He also remembers the grim postcards that came from the concentration camps, revealing that more and more friends of the family were falling prey to

Heinrich Poll
(1877–1939)

the Nazis. 'We knew that people were being killed', he said.

With time fast running out and the Nazi ferocity growing ever more rabid, Heinz Karger met Prof. Heinrich Poll whom the Nazis had forced out of the university in Hamburg. Prof. Poll considered it urgently important that all those scientists who were now forbidden to work, at least remain free to publish. Heinz Karger explained his position and recalls that Prof. Poll simply asked: 'Do you want to go to Switzerland?' The publisher's answer was just as simple: 'Yes!'

Prof. Poll had personal connections in Switzerland. Furthermore, Heinz Karger learned later that most of the medical faculty of the University of Basel were strongly in favor of having S. Karger in their city, even though another medical publisher, Benno Schwabe, was already based there. S. Karger was already known for its excellent books and specialized journals. Nothing comparable was being published in Switzerland in those days. The new publisher would be an asset to Basel. The new location would be a life-saver to the Kargers.

The old University of Basel

Paracelsus Theophrastus Bombastus von Hohenheim (1493–1541)
Portrait by Augustin Hirschvogel (1538)

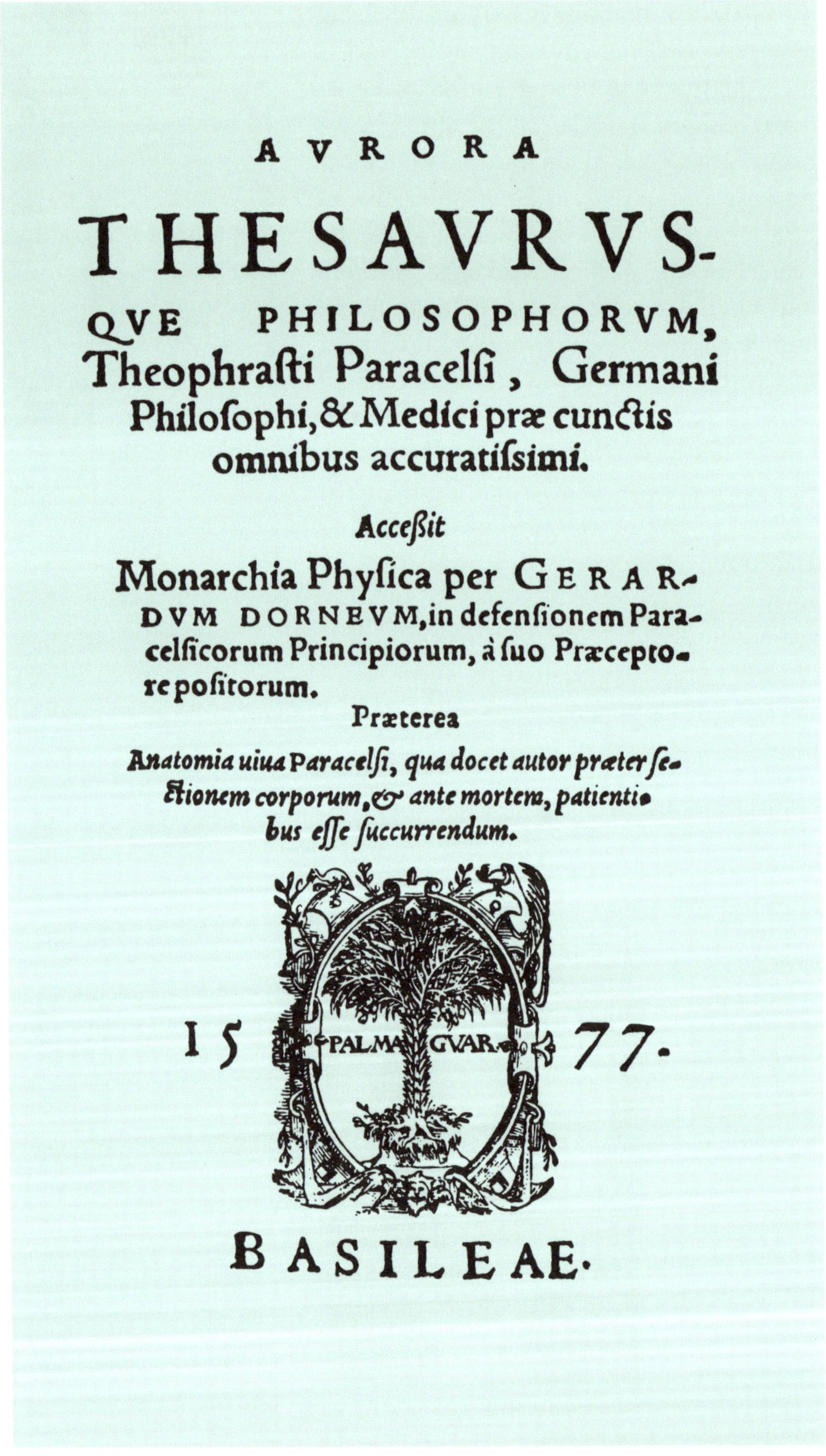

AVRORA

THESAVRVS-
QVE PHILOSOPHORVM,
Theophrasti Paracelsi, Germani Philosophi, & Medici præ cunctis omnibus accuratissimi.

Accessit

Monarchia Physica per GERARDVM DORNEVM, in defensionem Paracelsicorum Principiorum, à suo Præceptore positorum.

Præterea

Anatomia uiua Paracelsi, qua docet autor præter sectionem corporum, & ante mortem, patientibus esse succurrendum.

PALMA GVAR

15 77.

BASILEAE.

Title page of the 'Paracelsean' treatise in which a 'chemical-anatomical dissection' of the urine is recommended

Certainly, the long history of Basel suggested a congenial environment for such a high-quality book publisher. The art of papermaking in that city on the Rhine dates back to Heinrich Halbysen in 1433. Many mills were established in the old town district of St. Alban and the city became a famous center for the papermaker's art. Berthold Ruppel set up the first printing house in the city in 1468, some say 1466, having learned that revolutionary new art from Gutenberg himself. Basel was one of the first cities in Europe to which printing spread after Gutenberg opened his press in Mainz, less than two decades earlier. Basel was a significant trade and commercial center. The University of Basel itself was founded in 1460 and the intellectual life of the city was enriched by dealings with Erasmus of Rotterdam and lectures by the great, although highly controversial, medical pioneer Paracelsus.

Fritz Karger
(1903–1990)

But, more than 450 years later, the actual move of the publishing house was beset with difficulties despite the courageous help of its friends. Heinz Karger remembered with gratitude the help and advice of a publisher in Munich, Ernst Reinhardt, whose field was not medicine, but who had been a friend of his father's. Various friends helped persuade the German officials to extend the company's permission to publish in Berlin. But it was clear that these extensions were temporary and that the situation was steadily becoming more hazardous. The move to Basel had to be made, and quickly. Heinz Karger had considerable assistance from a lawyer who organized the legalities that were necessary to move the company. His younger half brother, Fritz Karger, who had been a lawyer in Berlin, had moved to Switzerland himself in 1936 and was entrusted with official representation of the company until the rest of the family could get there.

Ernst Reinhardt
(1872–1937)

They made the move in April 1937. Heinz Karger and his wife went by car, encouraged by the thought that Basel was just across the border, in sight of Germany. Young Thomas Karger, seven years old, went by night train with his grandmother.

'As young as I was, I quite realized what was going on', he recalls that perilous night journey. 'I was very much afraid.'

They escaped from Germany before the worst of the Nazi atrocities occurred and the company opened in Basel that spring with the same journals and the same books as it had in Berlin. The family believes that they were the only publishers to escape the Nazis without the loss of any publishing rights or any journals. According to the arrangement, the company's operations were to be allowed to continue in Berlin through the end of 1937 and that, thereafter, a Berlin office could be maintained, with its experienced staff, as a branch of the company in Basel. Such were the arrangements, but the reality proved far different. Just two weeks after the family escaped to Basel, the company's entire stock in Berlin was confiscated. The company, now safely in Switzerland, protested to the German authorities. The Nazi officials conceded that the agreed arrangements had all been legal, but explained, as Heinz Karger remembered the incident, that, in this case, 'the legalities could not be followed'.

The publishing house of S. Karger was not free of Nazi depredations, but both the company and the family were safe at last in a free country and there was hope. As Prof. Poll had told Heinz Karger when they were both living in the terror of Nazi Berlin, the German language needed a free environment and many authors were sure to seek it out.

First domicile of the S. Karger Company in Basel, Stapfelberg 2 (1937–1939)
A drawing by F. Reinhardt, Basel

2

War and Perseverance

ALLZEIT VORAN

S. K.

1890

With the move to Basel complete, Heinz Karger and his brother Fritz faced the agonizing problem of rescuing the stock and records of the company from the Nazis who played always by their own distorted rules of 'legality'. The next set of negotiations produced what looked like some progress. About half of the stock of books and journals was freed and the rest was shipped to a 'commissioner' in Leipzig for storage and eventual sale. The proceeds were supposed to go back to the Berlin Branch to finance operations there.

Meanwhile, the family had escaped from Germany with only a few thousand francs in cash. A boycott from the Nazi-controlled institutions was almost inevitable and the future looked most uncertain. There was in fact a Nazi interdict against German authors maintaining any contact with the new Swiss publishing house of S. Karger. Soon the Nazis also forced all German and Austrian editors to withdraw from the editorial boards of the Karger journals.

Nor were things altogether bright in the company's new home. Like every other place on earth, Basel had its own professional communities with their own deep-seated interests. Some of the local medical and publishing interests did not look kindly on serious new competition from foreigners. Furthermore, even though the transfer out of Germany was a profound relief, the family had deep roots in that country. The move was trying for everyone.

CH

Gotthelf School in Basel where Thomas Karger went to primary school

Thomas Karger still remembers the 'little nightmare' of a shy small boy in a strange city, a new school and even a language that was not too familiar. Swiss German is sufficiently different from the German of Berlin, so that he sometimes had a hard time understanding things. After the fashion of boys everywhere, some of his new Swiss schoolmates were rude and unwilling to be friends. But that problem cured itself in time and even the problems of the publishing house began to resolve themselves with the creative help of the hard-working Heinz Karger. He was known in the publishing world as a man of integrity with a strong knowledge of science and a dedication to humanitarian concerns. What he accomplished during a period of great trial showed that he was also an expert planner and an innovator of great talent.

The publishing house had been an institution of great strength in Berlin. Fortunately for the new enterprise, that strength lay mainly in its reputation for high quality and in the publisher's talent for sensing the importance of new fields of research. German institutes and clinics were forbidden to subscribe to Karger journals or buy Karger books. Sales and subscriptions dropped off alarmingly, but a few of the German institutions stubbornly ignored the prohibitions and many new subscriptions poured in from other European countries. Within a year and a half, the losses from Germany were counter-balanced and the journals had as many subscribers as they had during the Berlin period, sometimes more.

Swiss support for the move from Berlin had come first from the medical faculty of the University of Basel, particularly Profs. Arthur Brückner, Wilhelm Lutz and Rudolf Staehelin, who went to great efforts to give the idea a sympathetic hearing among the Swiss. A festschrift to Heinz Karger, published after the war, noted that the medical faculties of other Swiss universities joined the efforts.

Arthur Brückner
(1877–1975)

Wilhelm Lutz
(1888–1958)

Rudolf Staehelin
(1875–1943)

Under the energetic leadership of Heinz Karger, the company even founded some new journals. Among the first of these was *Cardiologia* [17], of which the first volume was published in 1937. This was designed as an international specialized journal to be published in four languages: German, French and Italian, the three official languages of Switzerland, and English. The journal is still published today as *Cardiology*. The 77th volume will be published in 1990. At the time of its origin, it served as a model for many future Karger journals.

The transformation into a truly international publishing house was dictated by many problems that had to be turned into opportunities. The publishing house was no longer German, and was forbidden contact with German scientists. The German language was no longer the preeminent language of science as it had been half a century earlier. There was opportunity in the fact that Nazi policies at home, and their aggressions throughout Europe, were leaving a vacuum of knowledge in medical science that needed to be filled. Heinz Karger decided it was time for the house to become an international publisher. The journals would be produced in Switzerland, mainly by Swiss editors, but would be truly international in scope and concept. He liked to use the Red Cross as an analogy, noting that it was an international organization, but used the Swiss flag in reverse colors as its symbol and the organization maintained extremely close ties to that traditionally neutral country. The titles of the Karger journals were changed from German to Latin in most cases. They now carried reports in German, English or French, sometimes Italian, at the choice of the authors. There were summaries in each of the other languages. This idea was not entirely new, but S. Karger put an unusually strong emphasis on the international character of their publications.

Gerhard Wolf-Heidegger (1910–1986)

In the postwar festschrift to Heinz Karger, Prof. Gerhard Wolf-Heidegger quoted the publisher as defining the role of *Acta Anatomica* [18], one of the new journals, as 'to promote the exchange of scientific thinking and to give researchers an opportunity to communicate their results to the largest possible audience ... in the interest of rebuilding a friendly better world'.

'Science does not have its headquarters with a single population', the publisher said in that dedication, 'but is the supranational common commodity of civilized humanity.' As the decade of the 1940s began, it must have been difficult to maintain faith in that splendid idea, but the history of the publishing house shows that the Kargers did so and that faith in the importance of their work probably helped the company survive.

During those years, the publishing house was recovering from the trauma of the abrupt move, but there must have been times when the safety of neutral Switzerland itself seemed to be hanging by a thread. The Nazi atrocities grew ever more ferocious during the 1930s. Only a year and a half after the Kargers left for Basel, the organized horror that the Nazis called the 'Reichskristallnacht' (Week of broken glass) swept across Germany. Businesses and homes were pillaged or burned. Many were injured or killed. Large numbers of Germans who abhorred the Nazi regime fled to Austria, or to Czechoslovakia only to see the Nazis take over those countries in quick succession. Many others had fled to France as one of the strongest bulwarks of democracy in Europe, but even that proved to be an illusion. Just two and a half years after the Karger family escaped from Germany, World War II began. Little more than a year after that, France fell. Denmark, Norway, Belgium and the Netherlands had already been overrun. Most of Europe was now in the hands of the Nazis or their collaborators. Switzerland could only hope to maintain its neutrality. There was no proof the Nazis would respect it.

50

It was a time of disruption and disaster for much of the world, and there was no place that could count itself entirely safe. That was the grim climate of world affairs when the medical publishing house of S. Karger celebrated its 50th anniversary. The event was marked in the pages of the journal *Dermatologica* [7], volume 81, by the editors of 12 Karger journals, most of whom were Swiss. The date was April 1, 1940. France was near defeat and the

‘Science does not have its headquarters with a single population, but is the supranational common commodity of civilized humanity.’

Heinz Karger (1895–1959)

killing was going on everywhere across the border only a few miles from Basel. Great Britain was bracing for an expected German invasion thrust. The Japanese were sweeping across much of Asia. The survival of human freedom in the world seemed much in doubt. Given that desperate climate, the editors' congratulations to the publishing house were as eloquent for what they omitted as for what they said.

Paying tribute to the high principles of the founder of the house, and to the idea of establishing a group of truly international journals, the editors added this comment: 'In the year 1937, the Company moved to Basle. With the assistance of the undersigned editors, the idea was carried still further forward to make the journals a sort of central organ for international cooperation in the domain of medical specialities. The journals are meant to serve as a channel for the continuous exchange of ideas in a manner similar to that made possible by the International Congresses.

Publisher's emblem
in use from 1937 to 1945

Après la mort, survenue en 1935, du fondateur de la Maison, et fidèle à l'esprit qui l'avait animé, son successeur fut amené, en 1937, à se fixer à Bâle, pour y établir, avec l'aide de divers rédacteurs, un centre de collaboration médicale internationale qui se révèle plus nécessaire aujourd'hui que jamais. A la liste des anciens périodiques transformés s'en ajouta de nouveaux; les uns et les autres se réjouissent d'un cercle de collaborateurs et de lecteurs de plus en plus étendu.

Les rédacteurs soussignés apportent ici à un éditeur sans cesse préoccupé de „servir" l'hommage de leur félicitations et de leur gratitude.

ANNALES PAEDIATRICI
(Jahrbuch für Kinderheilkunde)
(Fund. 1857) E. Freudenberg

DERMATOLOGICA
(Dermatologische Zeitschrift)
(Fund. 1893) W. Lutz

Monatsschrift für
GEBURTSHILFE UND GYNÄKOLOGIE
(Fund. 1894) E. Anderes, Th. Koller

GASTROENTEROLOGIA
(Archiv für Verdauungskrankheiten)
(Fund. 1895) A. Hurst, R. Staehelin

Monatsschrift für
PSYCHIATRIE UND NEUROLOGIE
(Fund. 1897) J. Klaesi

OPHTHALMOLOGICA
(Zeitschrift für Augenheilkunde)
(Fund. 1899) A. Brückner, H. Weve

PRACTICA
OTO-RHINO-LARYNGOLOGICA
(Fund. 1907) J. Berberich, E. Schlittler

RADIOLOGIA CLINICA
(Radiologische Rundschau)
(Fund. 1932) M. Lüdin

CARDIOLOGIA
(Fund. 1937) B. Kisch, W. Löffler

BIO-MORPHOSIS
(Fund. 1938) H. Bluntschli

CONFINIA NEUROLOGICA
(Fund. 1938) E. A. Spiegel

Schweizerische Zeitschrift für
ALLGEMEINE PATHOLOGIE UND BAKTERIOLOGIE
(Fund. 1938)
A. v. Albertini, A. Grumbach, H. Mooser

The publishers and editors are convinced that such cooperation on a wide international basis is more than ever essential today. Therefore, not only were the old journals given a new form, but new international organs have been called into being as well.'

The editors' only reference to the world disaster that was unfolding everywhere beyond the Swiss border was this statement: 'In spite of the unfavorable circumstances obtaining, these periodicals made their way in a short space of time and the many personal acknowledgements received from scientists all over the world may well be proof that we are on the right road.'

In fact, the lamps of science and human freedom were being snuffed out all over Europe. No one could be sure how long it would be before Nazi panzer divisions would bring 'the unfavorable circumstances' across the border into Switzerland itself.

Even without that disaster, Switzerland was almost entirely isolated now from the rest of the world, and isolation is deadly to an international publisher. There was no reliable way to get journals and books to buyers, and great difficulty in getting manuscripts from non-Swiss authors to Basel.

Just before the war began, the publishing house had moved into small and frugal quarters on the Holbeinstrasse. It proved to be a wise choice because the offices were small and easy to heat during the winters, when the war made fuel more and more scarce. But

Second domicile of the S. Karger Company, Holbeinstrasse 22, Basel (from 1939 to 1953)

Markus Guggenheim
(1885–1970)

Emil C. Barell
(1874–1953)

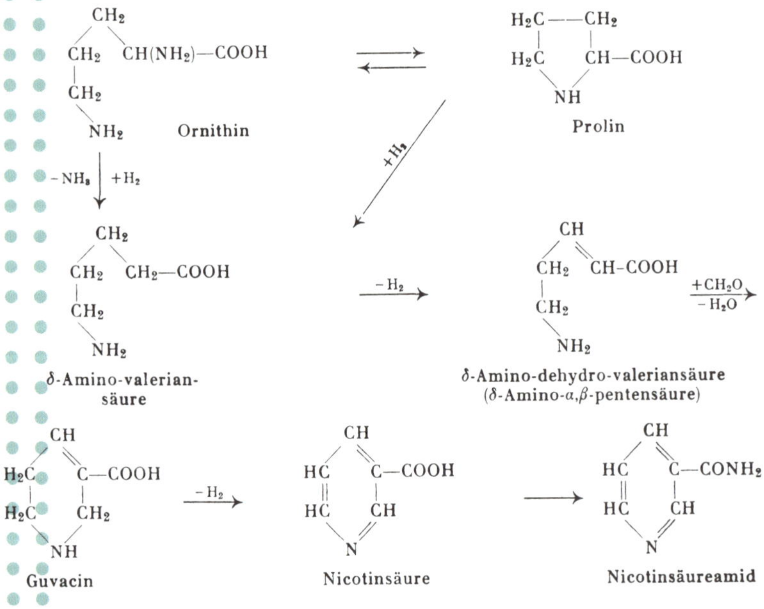

even with such careful economies, it soon became obvious that the company could no longer survive on its own resources. At that point, three powerful Swiss resources came to the rescue.

At the suggestion of Dr. Markus Guggenheim, who would become one of the first Swiss authors of S. Karger, with his authoritative book, *Die biogenen Amine* [19], Heinz Karger went to see the General Director of Hoffmann-La Roche, Dr. Emil Barell. Luckily, Dr. Guggenheim, then head of research of Hoffmann-La Roche, knew and appreciated S. Karger from the days in Berlin: he supported the request, and Heinz Karger received a friendly and constructive response. The director said he would make some funds available through a foundation and that the money need not be repaid. The company was prepared to do this if the other main industrial powers in Basel, Ciba and Sandoz, would also contribute. The arrangement was intended to continue so long as the war emergency lasted, a remarkably open-ended piece of generosity when viewed from the state of the world in 1940. This too was a tribute to the Karger reputation. The pharmaceutical companies were convinced that the publisher represented a resource for medicine and science that was worth considerable expense to preserve.

CIBA

Chemische Fabrik vormals Sandoz, Basel

ROCHE

The three companies agreed on the proposition and S. Karger again had a chance to survive and even to continue its contributions to science and medicine. It had a chance, that is, if only some way could be found to get the volumes to the doctors and scientists who wanted and needed them.

Heinz and Fritz Karger had made a concerted effort to build up international contacts while this was still possible, but in the fall of 1939, Poland had been overrun and, after the summer of 1940, most of Europe was in Nazi hands. Nevertheless, the publishing house continued to produce its journals without interruption. Even after the occupation, it proved possible to send manuscripts and proofs to and from the Netherlands without much trouble, and individual issues of the journals could often be sent out too. But the publishers soon learned that larger packages attracted attention and were intercepted.

Fritz Karger remembered one case in particular in which a work passed to and from the Netherlands in manuscript and proof form three times with no notice from the Nazi censors. Yet, when the publisher sent the author his copies of the published version, the package was confiscated and destroyed on the incredible grounds of being 'Bolshevik'. Other packages of medical and

scientific texts were regularly intercepted and destroyed as 'staatsfeindliches Schrifttum' (subversive literature). There were many such experiences and the Kargers finally stopped trying to send copies to authors and subscribers. Publication still continued, but the company began storing all issues in the cellar below the offices.

Members of the family and employees long remembered how the books and journals began to pile up, first to fill the cellar and then the corridors between rooms, and finally every conceivable nook and cranny available. Heinz Karger's decision to keep publishing was courageous, but difficult to live with. 'We didn't question his decision', one employee remembered long afterward. 'We just found it hard to squeeze around all those stacks.'

That growing glacier of printed material was a strange, but potent testimony of faith in the survival of freedom. The Kargers continued to publish, not just minimal numbers of volumes 'for the record', but in large quantity. Heinz Karger was clearly convinced that the war would end someday and that the Nazis would be defeated so that scientific discourse could be resumed with no fear that it would be burned as 'enemy to the state'.

The continued publication also gave the Swiss a medium for communication with the rest of the world unhampered by Nazi censors. Even for doctors and scientists of the occupied countries, the survival of the publishing house must have seemed a ray of hope as their only possible connection and avenue of communication with the free world of science. As Fritz Karger recalled many years later, however, the moral success was great, but financially there was nothing.

Even the stock that had been stored in Leipzig was finally lost. A man who worked for the company in Basel decided to resign for personal reasons. His name had been listed in the German address list of booksellers as the director of the publishing house. Whether he was listed for the whole company

or just the Berlin branch is no longer known, but the Nazis used his resignation as an excuse to take over all the remaining Karger assets that they could grab. The remaining stock that had been stored in Leipzig for several years was seized. Since these were now Swiss, not German, assets, the company in Switzerland protested and kept up the fight for two years. In the midst of the war, the Nazis were even less concerned with legality than they had been before, and the efforts to rescue the stock in Leipzig failed. In the end, virtually everything was destroyed, including the stored journal volumes. So far as the family could learn, the only surviving stock was 200 copies of a history of medicine. The Kargers gave these copies to the heirs of the author, who had died years before, and the books were stored in a Leipzig bookstore. A week later, the bookstore and all 200 copies were destroyed in an air raid.

In Switzerland, the company continued its desperate struggle to survive and to maintain whatever contacts were still possible with scientists abroad. Every month it seemed that the task was more difficult and the obstacles greater. Then, in the summer of 1944, all the publisher's contacts with countries beyond Switzerland were broken. But, by that time, it was clear that the Nazis were being defeated and that the years of trial and isolation must be coming to a close. The printing went on and the offices on the Holbeinstrasse became more and more jammed with undistributed volumes.

Then the Allied armies swept through France and the rest of Europe and the war was over at last. The Nazi plague that had beset the entire continent since the early 1930s was finally ended, and civilization could set

The grounds of the Leipzig Technical Fair after the 1945 air raids

about rebuilding. The journals and books that had been accumulating so greatly in the offices of S. Karger could be sent to their intended recipients with no fear that they would be intercepted or censored. The three industrial companies which had helped the publisher survive during the war years had been receiving free copies of the Karger volumes. The companies were now able to send these out too as gifts to physicians and collaborators throughout the world.

The Karger publications helped the world of science restore the continuity of work that had been so gravely disrupted by the war. That effort was well under way even before the last guns had fallen silent.

The importance of the continuity and the survival of the free expression of ideas was stated eloquently by three scientists writing an introduction to the new Karger journal *Acta Anatomica* [18] in the summer of 1945. The war had just ended in Europe and was about to end throughout the rest of the world.

'Science has also suffered severely during a war of devastation: scientists have been taken from their work, technical and literary resources destroyed, and, not least in importance, there has been an ever-spreading increase in power of forces which have nothing to do with the objective search for truth.

The war is now finished. The danger for the freedom of research is over, and science looks forward to a period of undisturbed development.

Now is the time for reconstruction; and science, which is not the property of any one individual, or nation, but a common heritage of the entire civilized humanity, is above all other fields of human endeavour called upon to show the importance and necessity for international co-operation. It is of first importance to assist the interchange of scientific ideas and to give research workers the opportunity to bring the results of their work before as large a number of their colleagues as possible.'

That was their stated objective in establishing a new journal to publish a wide range of works in anatomy, histology, embryology and cytology, all of them subjects that were to undergo incredible advances in the decades ahead. The advances were to have many beneficial consequences for the understanding of the human system and the better treatment of human disease.

The editors of the new journal concluded their introduction with this paragraph: 'It is the wish and hope of the editors that *Acta Anatomica*, as a modest contribution to the building up of a better and more peaceful world, will be of service in encouraging international scientific co-operation. In this spirit the undersigned confidently offer the first number of the journal to their colleagues.'

It was signed by Gösta Glimstedt, of Lund, Sweden; Tibor Péterfi, of Istanbul, Turkey, and Gerhard Wolf-Heidegger, of Basel.

In 1946, only a year after the end of the war, Heinz Karger was able to go back to the chemical companies that had helped him survive and inform them that the subsidies were no longer needed. The publishing company was independent and self-supporting again and the faith the industrial officers had demonstrated during the darkest years of the 1940s had been justified. An official of one of the companies told the publisher that the occasion was one of the very few in his life in which anyone had actually wanted to give up financial support.

But S. Karger wanted independence, and their stature as an international publisher now seemed assured. The time had come at last, not just to survive, but to rebuild.

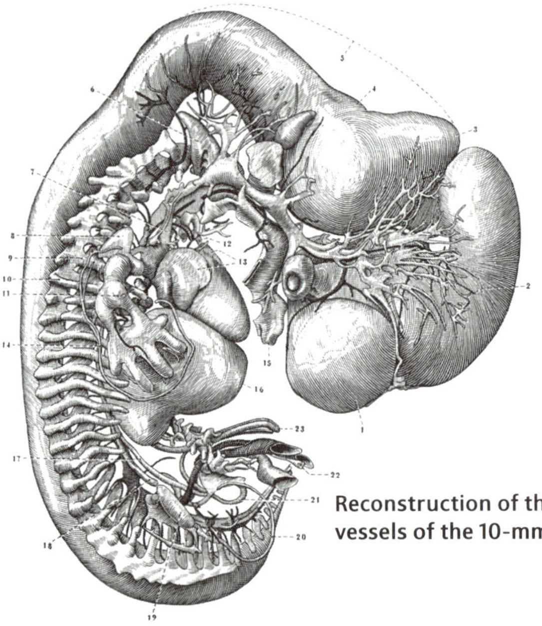

Reconstruction of the larger vessels of the 10-mm embryo

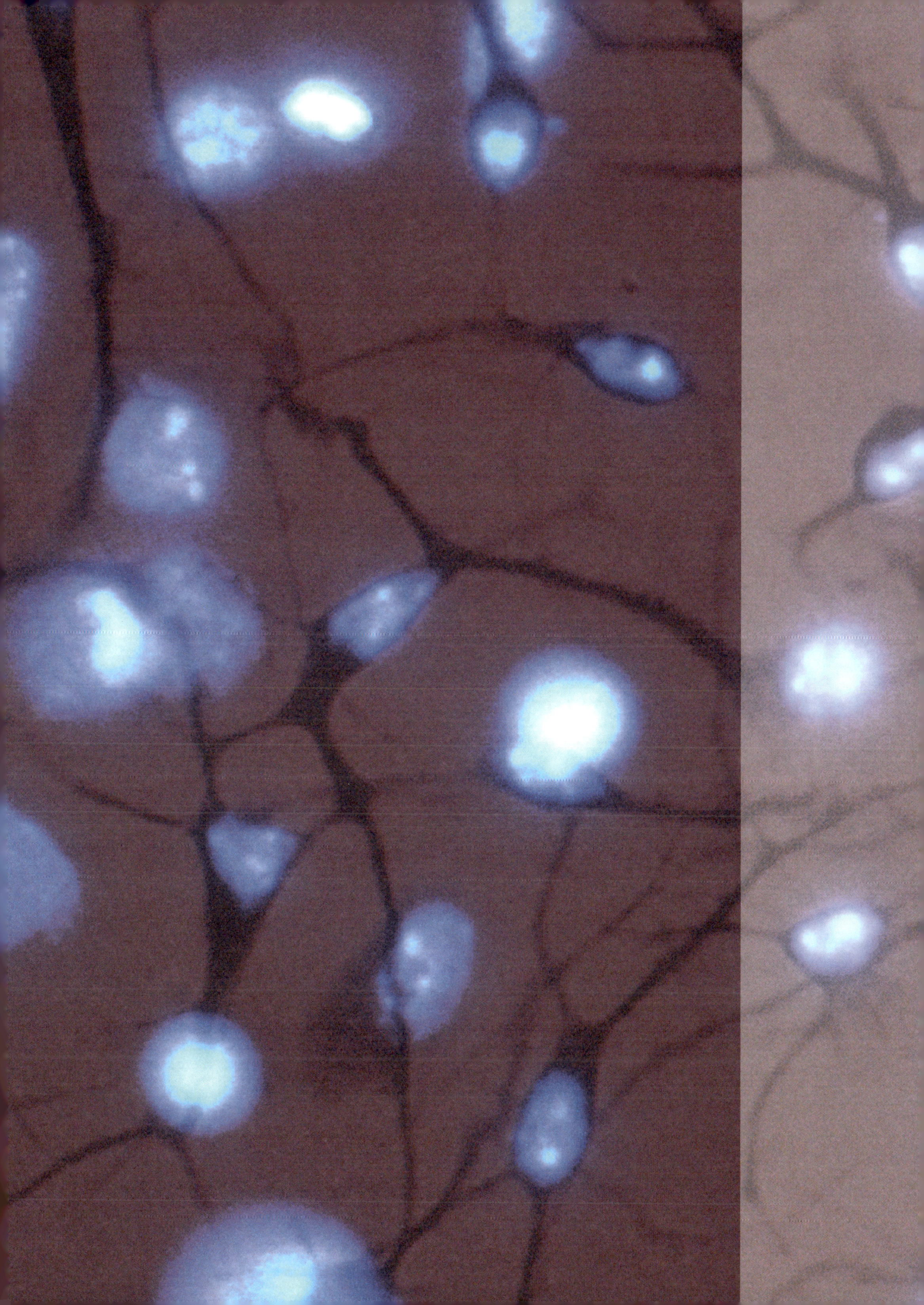

3

Rebuilding in Peace

In spite of the war's devastation, it took only a few years for science to come into its own again as one of the main driving forces of 20th-century civilization. In 1945, S. Karger was in an excellent position to help that resurgence. The books and journals that had been choking every available bit of space in the publisher's offices for years, were mailed out. Authors and readers alike were delighted to see the continuity of science made manifest in those volumes that had been lying fallow so to speak in Basel.

That thought was expressed by Dr. Paul Kallós of Helsingborg, Sweden, in renewing the work of a Karger publication of which he had edited the first volume. This was *Progress in Allergy* – Fortschritte der Allergielehre [20]. It had been published just before the outbreak of the war in Europe and was out of print long before the war ended: 'The reception which volume 1 then received encouraged me to continue this work as soon as conditions would permit an international co-operation', Dr. Kallós wrote in a foreword to the series after the war. 'The quick development in the field of allergy and still more its theoretical and practical importance made this continuation all the more necessary. Owing to the fact that during the years of hostilities the European Continent was practically isolated, I deemed it advisable to endeavor to collect and review the most important results and perspectives.

'It is to be hoped that this volume, which will soon be followed by others, will in some fields be a stimulating help to the research worker and a valuable guide to the clinician.' The editor expressed his thanks and apprecia-

Paul Kallós
(1902–1988)

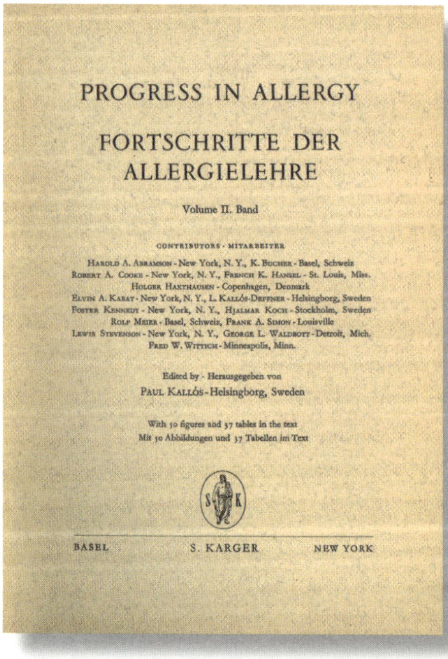

PROGRESS IN ALLERGY

FORTSCHRITTE DER ALLERGIELEHRE

Volume II. Band

CONTRIBUTORS · MITARBEITER

Harold A. Abramson - New York, N. Y., K. Bucher - Basel, Schweiz
Robert A. Cooke - New York, N. Y., French K. Hansel - St. Louis, Miss.
Holger Haxthausen - Copenhagen, Denmark
Elvin A. Kabat - New York, N. Y., L. Kallós-Deffner - Helsingborg, Sweden
Foster Kennedy - New York, N. Y., Hjalmar Koch - Stockholm, Sweden
Rolf Meier - Basel, Schweiz, Frank A. Simon - Louisville
Lewis Stevenson - New York, N. Y., George L. Waldbott - Detroit, Mich.
Fred W. Wittich - Minneapolis, Minn.

Edited by · Herausgegeben von
Paul Kallós - Helsingborg, Sweden

With 50 figures and 37 tables in the text
Mit 50 Abbildungen und 37 Tabellen im Text

BASEL S. KARGER NEW YORK

Heinz Karger in 1955

tion to contributing scientists in the United States and Sweden and to Heinz Karger 'for his generous support and kindness'.

Allergy, and its underlying science, immunology, has been one of the key areas of medical and scientific advance ever since the war and has been one of the prime areas of specialization by Karger books and journals. Dr. Kallós was an editor of Karger publications until his death in 1988.

In immunology, as in many other important fields, the period that began in the late 1940s and early 1950s was far more than a resumption of scientific business-as-usual after nearly a decade of disruption. Instead, a new era was being born, even more revolutionary than the years before World War I that had seemed such a fertile opportunity for scientific publishing to Samuel Karger.

In the new era it was to be the son, Heinz Karger, who would first seize the opportunities that a new generation of medical and biological research workers was creating. Those contributions of science were to expand and grow far beyond what most scientists would

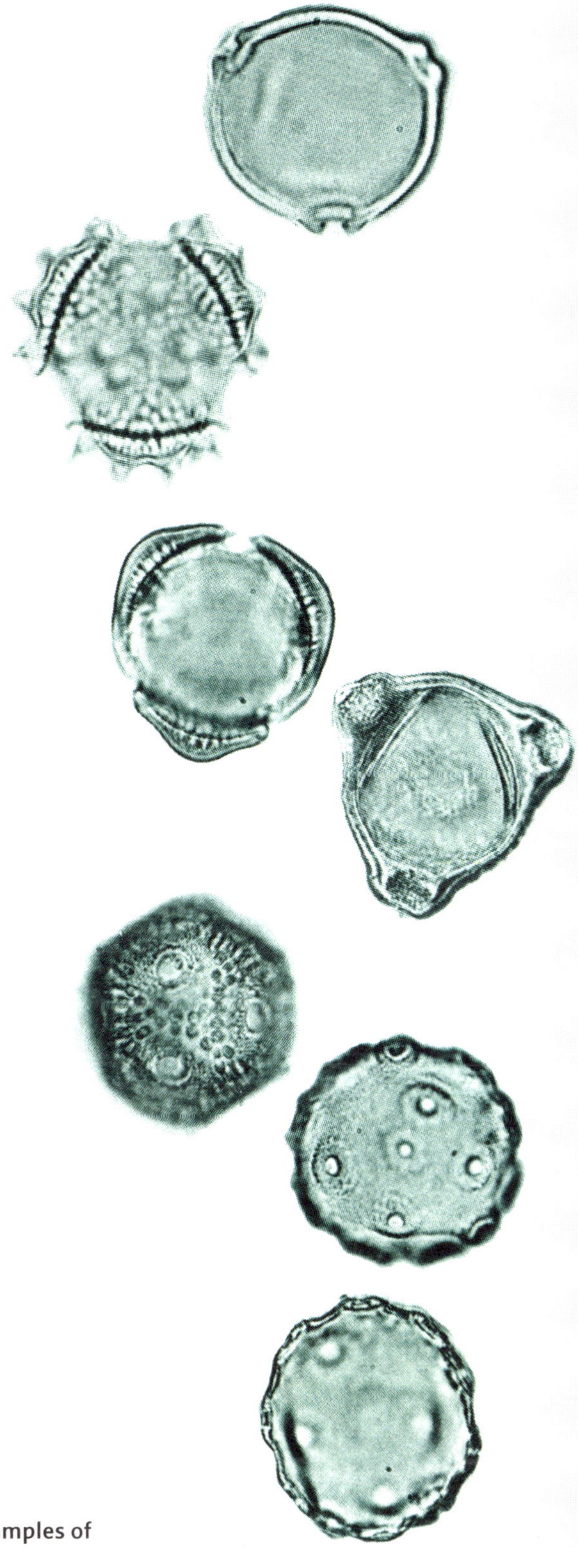

Samples of angiosperm pollen

have dared to predict in their wildest imaginings in the mid-1940s. Indeed, throughout the world, an immense flowering of many fields of biomedical science was just beginning.

'All over Europe an increasing demand for the latest scientific literature of the free world had to be met whilst on the other hand the medical sciences developed with a rapidity formerly unknown', said a tribute to Heinz Karger written many years later by a bookseller in Amsterdam. 'Still, Karger would not have been Karger if he had not carefully maintained the high standards of his publications.' Included in the new ventures were virology, immunology, cytogenetics and all the varied disciplines that are now known under the general name of molecular biology. It seems clear that Heinz Karger kept in close touch with these developments because the company was soon playing an important role in publishing the progress of many of the most active fields. Furthermore his company was increasingly broadening its horizons beyond medicine.

Only three years after the fighting stopped in Europe, the first volume of Theilheimer's *Synthetic Methods of Organic Chemistry* [21] was published and new volumes appeared in succeeding years, putting the publisher solidly in that major branch of chemistry. In 1952, the company published the *Proceedings of the First International Congress for Allergy* [22], a huge work of 1,175 pages. This publication of the proceedings of an international congress was the beginning of a new specialty of the publishing house. That specialization in publishing such proceedings was later extended to the records of important conferences in many fields of medical science.

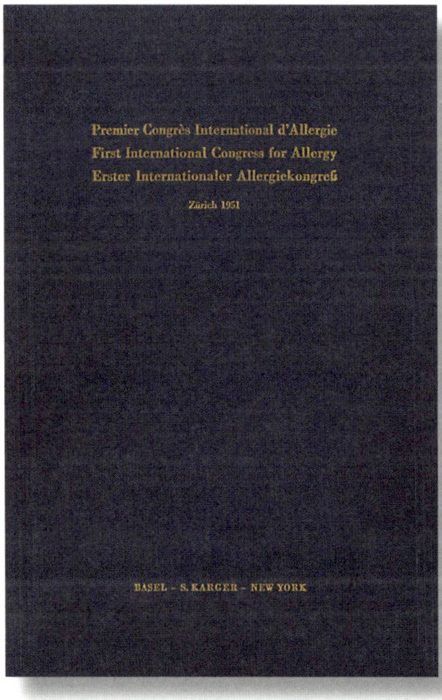

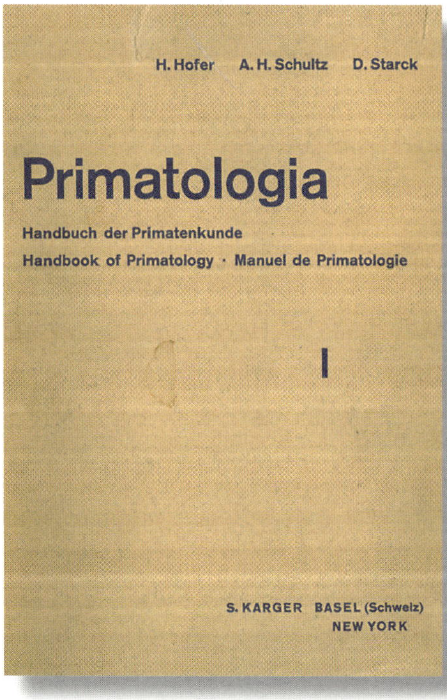

In 1956, only a decade into the postwar era, S. Karger published the first volume of *Primatologia – Handbook of Primatology* [23]: the beginnings, for the company, in that field too. Within a few years, the publisher had grown substantially larger than it was during the best of the Berlin years. In addition to the seven journals brought from Germany, another 20 were added. S. Karger also began a number of serials, yearbooks and 'Progress' volumes in special fields, among them the highly important research realm of virology.

Heinz Karger's contributions to medical and scientific publishing were appreciated in their home community as well as in the international world of science. In 1948, a little more than 10 years after they came to Switzerland, both Heinz Karger and his brother Fritz became Swiss citizens. In those days, this was an extremely short time, and Heinz Karger was very proud of this. In 1955 the medical faculty of Basel University awarded him an honorary doctor of medicine degree in recognition of everything he had accomplished since those desperate days in which the family and the company first arrived from Berlin. He already had a doctoral degree in economics, but was particularly pleased by the university's action because he was not a native Basler and therefore was less likely to be given such an honor.

The university's links to fields in which S. Karger has made important contributions go remarkably far back in the history of modern science. One case exemplifies this and also underscores the truly international nature of modern science, a concept that has guided the publishing house since it first took root in Basel.

Heinz Karger after having been awarded the honorary doctorate in the Martinskirche Basel (talking to Markus Guggenheim and his wife)

Publisher's emblem in use from 1946 to 1976

In 1872, a chemist, Friedrich Miescher, was appointed professor of physiology at the University of Basel and became noted for his studies of the spermatozoa of salmon, frogs and other species. His place in scientific history today rests on the fact that he discovered the substance now known as DNA. It was a little appreciated achievement at the time. Indeed, the scientist himself never fully realized the importance of what he had discovered. But, today Basel has a research institute named in his honor.

It was not until the wartime 1940s, however, that American scientists led by Oswald Avery discovered the crucial role of DNA in heredity. Then, in 1953, James D. Watson and Francis H.C. Crick, an American and an Englishman working at the University of Cambridge, discovered the structure and mode of replication of the DNA molecule. That discovery set the stage for the revolution in molecular genetics that is still unfolding at an ever accelerating pace today. In 1989, S. Karger published the latest in a series of proceedings of international gene mapping workshops [24]. The huge volume, completed within a few months of the close of the conference, at Yale University in New Haven, Conn., set forth the latest advances in a field that rests ultimately on the understanding of nucleic acids that began to unfold a century ago. Long before the first gene mapping workshop, S. Karger was making an important contribution to that crucial branch of science and to the men and women who have been its pioneers.

Friedrich Miescher
(1844–1895)

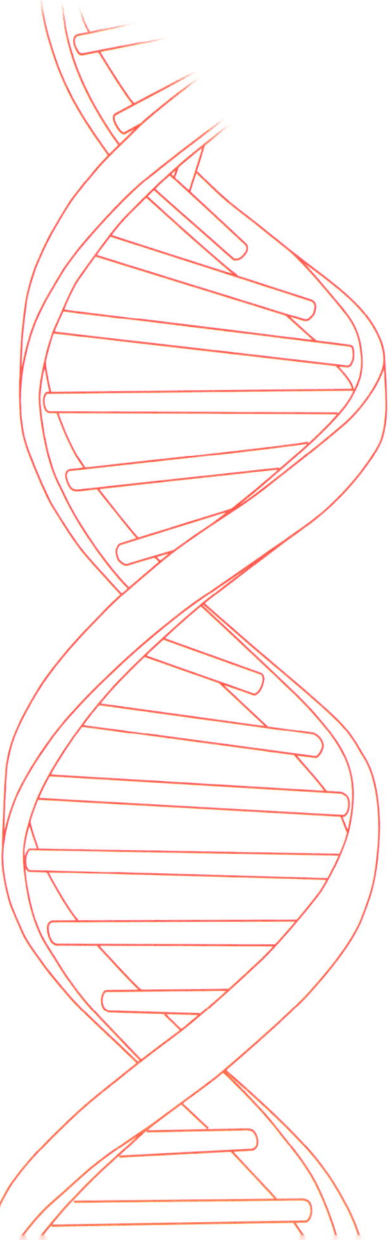

Harold P. Klinger
(1929–2004)

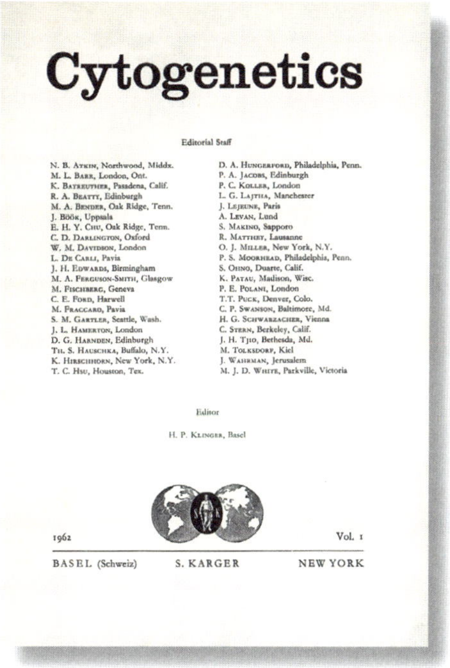

Cytogenetics

Editorial Staff

N. B. Atkin, Northwood, Middx.
M. L. Barr, London, Ont.
K. Bayreuther, Pasadena, Calif.
R. A. Beatty, Edinburgh
M. A. Bender, Oak Ridge, Tenn.
J. Böök, Uppsala
E. H. Y. Chu, Oak Ridge, Tenn.
C. D. Darlington, Oxford
W. M. Davidson, London
L. De Carli, Pavia
J. H. Edwards, Birmingham
M. A. Ferguson-Smith, Glasgow
M. Fischberg, Geneva
C. E. Ford, Harwell
M. Fraccaro, Pavia
S. M. Gartler, Seattle, Wash.
J. L. Hamerton, London
D. G. Harnden, Edinburgh
Th. S. Hauschka, Buffalo, N.Y.
K. Hirschhorn, New York, N.Y.
T. C. Hsu, Houston, Tex.
D. A. Hungerford, Philadelphia, Penn.
P. A. Jacobs, Edinburgh
P. C. Koller, London
L. G. Lajtha, Manchester
J. Lejeune, Paris
A. Levan, Lund
S. Makino, Sapporo
R. Matthey, Lausanne
O. J. Miller, New York, N.Y.
P. S. Moorhead, Philadelphia, Penn.
S. Ohno, Duarte, Calif.
K. Patau, Madison, Wisc.
P. E. Polani, London
T.T. Puck, Denver, Colo.
C. P. Swanson, Baltimore, Md.
H. G. Schwarzacher, Vienna
C. Stern, Berkeley, Calif.
J. H. Tjio, Bethesda, Md.
M. Tolksdorf, Kiel
J. Wahrman, Jerusalem
M. J. D. White, Parkville, Victoria

Editor

H. P. Klinger, Basel

1962 Vol. 1

BASEL (Schweiz) S. KARGER NEW YORK

Dr. Harold P. Klinger, now of Albert Einstein College of Medicine in New York City, recalls how association with the publishing house influenced his career. He describes himself as one of those 'lucky to have drifted into genetics' in the 1950s when the modern wave of discoveries was just beginning. He notes, for example, that even something as fundamental as the correct number of human chromosomes was not established until 1956. Before that year, it had been assumed, on crude and grossly misleading evidence, that the correct number was 48. The actual number is now known to be 46 and the structure and function of many genes of each of those chromosomes is known.

In the 1950s and early 1960s, reports that linked serious human disorders to chromosome abnormalities, particularly abnormalities of the sex chromosomes, were greeted with much skepticism. In fact, there was a lack of real experts in the biology of the chromosomes. The chemistry of genetics was a field still in its early gestation period. Young research workers were becoming increasingly frustrated by the shortage of competent reviewers in their new field. The lack of reviewers made editorial decisions by many scientific journals haphazard and often ill informed. They were not yet equipped to gauge the importance of reports submitted for publication in the unfamiliar new field of cytogenetics.

Dr. Klinger, an American who was studying and doing research at the University of Basel, was one of those frustrated young scientists. He and his wife were about to go on a short vacation in Italy, when he received a rejection slip from a prestigious journal that had decided it would not print a report on a newly discovered human chromosome abnormality of which he and his co-authors were justly proud. The journal editors told him that, in the expert opinion of their reviewers, the discovery that they found so new and exciting was probably just an artifact of optical distortion. It is not the kind of judgement that one likes to take on vacation. But, under the serene influence of the Italian seaside, Dr. Klinger came to an important conclusion: if there were no adequate journals or reviewers for this new speciality in genetics, why not simply start a new journal with a panel of competent reviewers? It was certainly a logical idea, but also a little frightening to a young scientist early in his career. Returning to Basel after the vacation trip, he suggested the journal idea to Dr. Wolf-Heidegger, who was chairman of the Department of Anatomy at Basel University, as well as editor in chief of the Karger journal *Acta Anatomica* [18]. Dr. Wolf-Heidegger had been extremely helpful and encouraging to Dr. Klinger who had only recently received his MD degree and was working on his PhD at the university. But the young doctor was very much worried that the older scientist might consider him presumptuous in thinking that he was experienced enough to found an entirely new scientific journal.

Far from being offended, Dr. Wolf-Heidegger arranged an appointment for him with Heinz Karger.

'I feared embarrassment and humiliation when I met with him', Dr. Klinger wrote many years later, 'because at that time editors in chief of European journals were all well-established men of distinction, heads of university departments and institutes, who ruled like kings'.

Heinz Karger, who struck the young American as a fatherly sort of man, listened politely to the presentation and, at the end, was silent for perhaps five seconds. Then, to his amazement, the publisher asked simply: 'How much general expense money do you need to start?'

'I value their word far more than I would lengthy contracts in fine print.'

Young Dr. Klinger was startled and thrown off balance by the response. He had not yet even considered the financial aspects of his project. Still, he felt compelled to play the man of affairs and give as rapid and concise an answer as Heinz Karger had just offered him. Groping hurriedly for an appropriate number, he picked one that seemed to him major – his own monthly salary. Dr. Klinger said he thought about 200 Swiss Francs would be needed. It was then the equivalent of about US $50.

'To my relief', Dr. Klinger recalls, 'he gave me a fatherly smile, wrote a check and handed it to me.'

Still playing the worldly man of affairs, he thanked the publisher and put the check in his pocket without looking at it. When he did take it out later, the young would-be journal founder discovered that Heinz Karger had added another zero to the estimate and had made the check out for 2,000 Swiss Francs. Neither then nor at any other time in the genesis of the journal did Dr. Klinger sign any contract with S. Karger. That journal, still highly successful today, was one of the

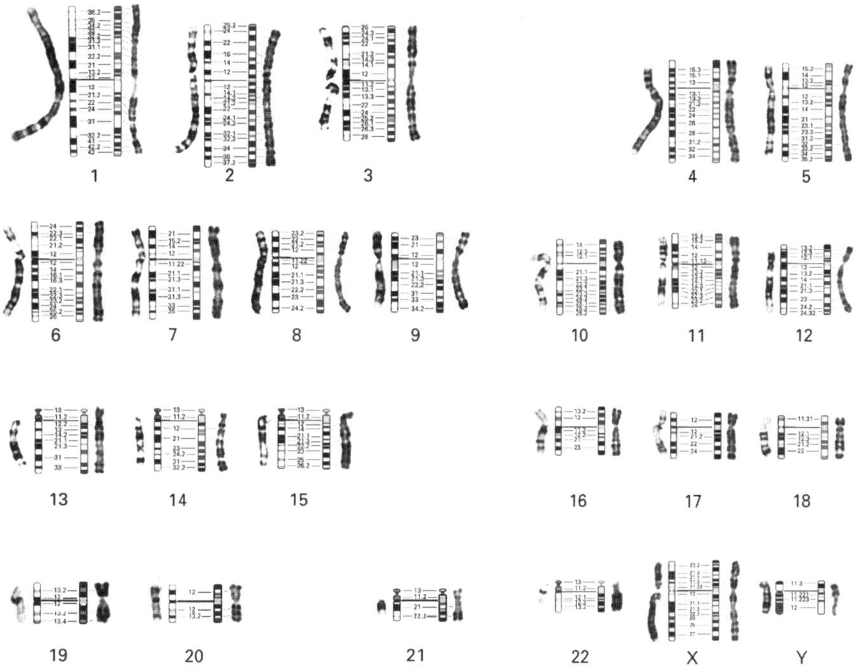

The Normal Human Karyotype G- and R-bands

last that Heinz Karger lived to found. The informality and simplicity with which it was arranged still impress Dr. Klinger: 'This quick start of the journal, uncomplicated by a single written word, only a hand-shake, is characteristic of my experience with the Kargers', says Dr. Klinger. 'I value their word far more than I would lengthy contracts in fine print.'

The first volume of the journal, then called *Cytogenetics* [25], was published in 1962, with Dr. Klinger as editor and an editorial roster full of names that have become famous during the past three decades. The periodical is now called *Cytogenetics and Cell Genetics* [25], and Dr. Klinger is still its editor in chief.

One of Heinz Karger's great contributions to the publishing enterprise was his determination that the books and journals be international in scope. He traveled regularly through Europe, and beginning in 1947, to the United States and Canada as well. The objectives were to keep up to date on world science, to talk to authors and editors and to make sure that all publications from S. Karger were available – and read – in the major libraries.

Many German scientists, who had been closely associated with the publishing house while it was still in Berlin, were now in America. The publisher traveled extensively through the United States and Canada renewing old friendships and making new contacts.

Friends and colleagues in the publishing business often cited his great sense of responsibility to publish expensive books when scientists convinced him of the importance of the works. In this context, one colleague cited a splendidly produced atlas of human anatomy by Dr. Wolf-Heidegger, of which a completely revised edition is being published on the occasion of S. Karger's centenary [26]; a work on the thalamus of the rhesus monkey by J. Olszewski [27] and another on the cytoarchitecture of the human brainstem by J. Olszewski and D. Baxter [28]. Two journal editors, Heinrich Heusser of Basel and Peter

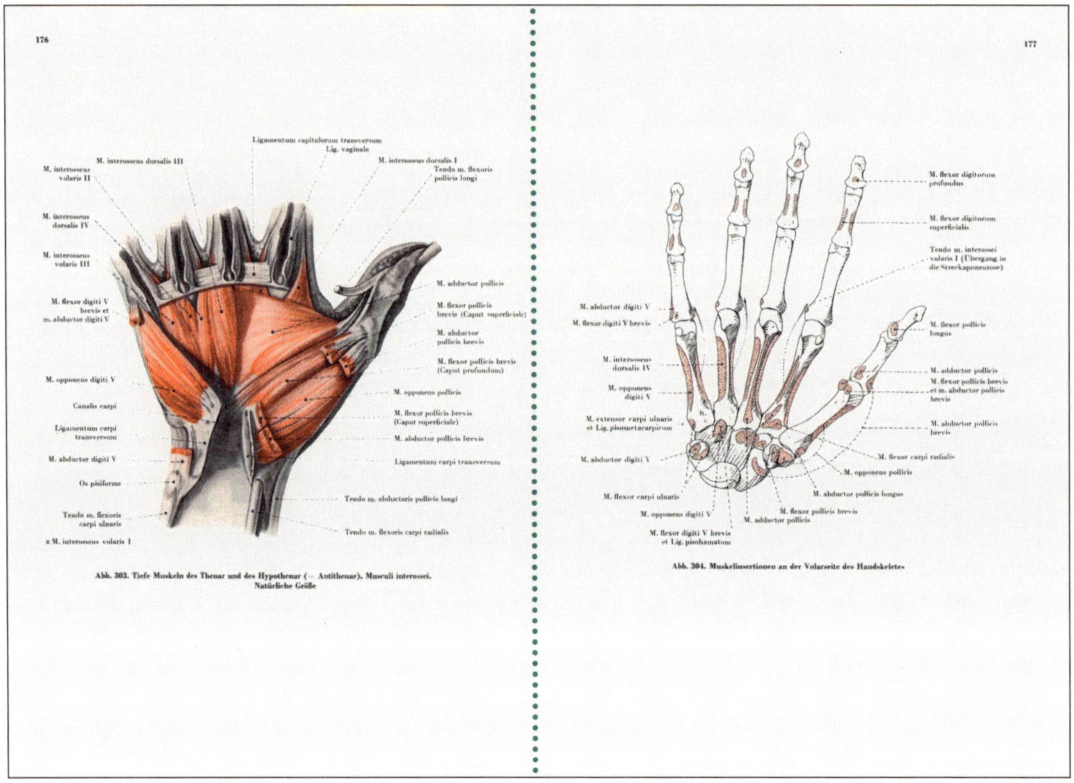

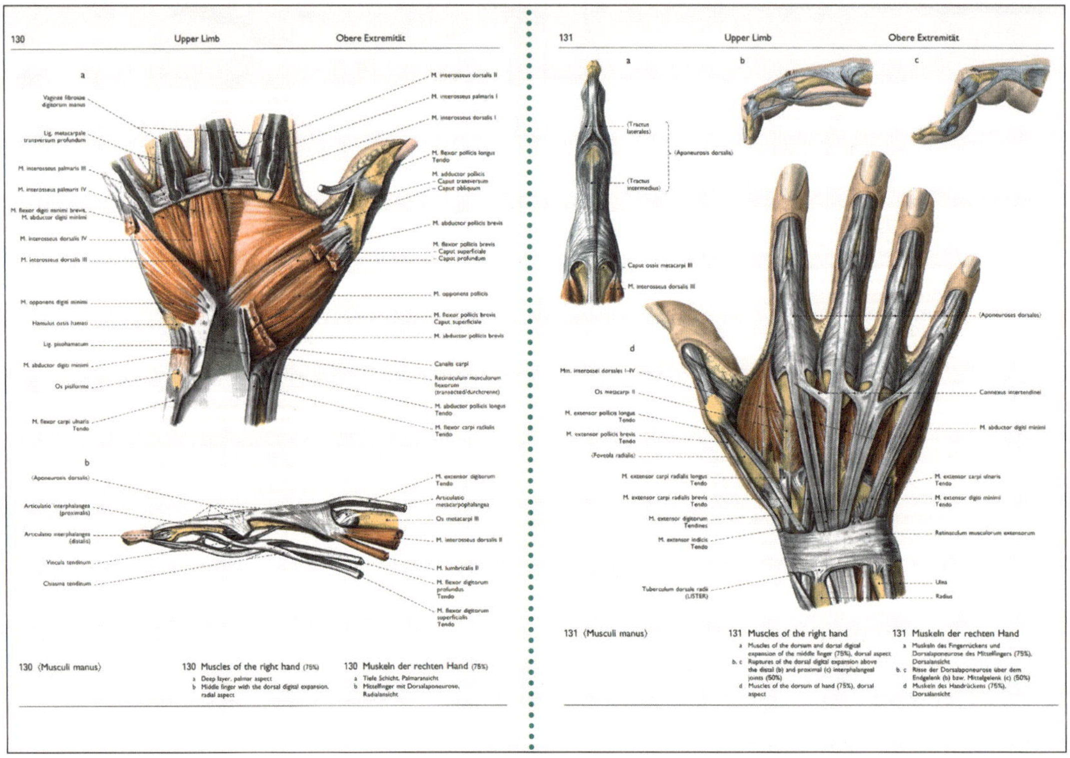

Illustrations from the 1954 edition of the *Atlas of Systematic Human Anatomy* by Wolf-Heidegger in comparison with the completely revised edition 2005

A. Narath of New York, described him as having 'an astonishing knowledge of the needs of medical science'. Others noted that his interests were not limited to strictly medical concerns: 'Karger does not restrict himself to medicine in a limited sense', one colleague said of him. 'Thanks to his reputation, other scientists have applied to him too.'

Less than a decade after the war, the company had already outgrown the character of a small, family-run publisher of high-quality books and journals in the German language. In the early 1950s the publishing house had nearly 1,000 books in print and was publishing 18 international journals.

In a conversation only a year before his death, Heinz Karger expressed great satisfaction that he had been able to show his early supporters in Switzerland that the publishing house had emerged into the postwar world as an institution of global importance. That importance continued and expanded throughout his life.

'His was a proud, traditional institution, very proud of the product they produced, scientifically, culturally and esthetically', said Dr. Joseph L. Melnick, of Baylor College of Medicine, Houston, Tex., an internationally known American virologist. Dr. Melnick first became editor of a Karger journal, during the mid-1950s, when virology was just beginning the revolutionary growth that has characterized the field ever since. He said the volumes produced by S. Karger in those days looked like elegant library volumes of the 19th or early 20th century. 'They believed that a book ought to look like a book and read like a book', he said.

'They believed that a book ought to look like a book and read like a book.'

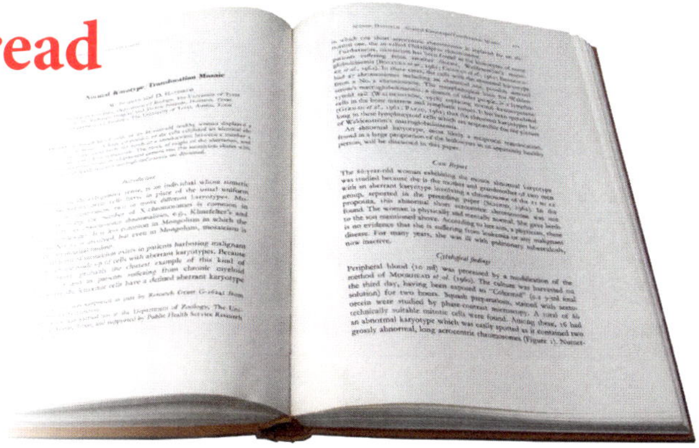

While maintaining that quality, Heinz Karger was shifting more and more from multilingual publication to reliance on English. Ever since the war, he had recognized the United States as the major new power in medicine and science. English was clearly the modern successor to German as the international language of science. He had saved his company from the Nazis in the 1930s and, two decades later had just about completed its transformation to a major international publisher, well prepared to play a role in the incredible flowering of science of the last half of the 20th century. But, at that point, he died.

Heinz Karger's death came suddenly in the spring of 1959 while he was driving with friends in Switzerland. He was only 64 years old. His son Thomas, who took over the publishing house immediately, believes the hard years of conflict with the Nazis probably shortened his father's life. But the man who rescued his family and his company from tyranny in the 1930s long outlived that tyranny and those who inflicted it on the world. Heinz Karger left a monument in the form of a publishing house dedicated to progress in science and the healing arts and to the free exchange of ideas throughout the world.

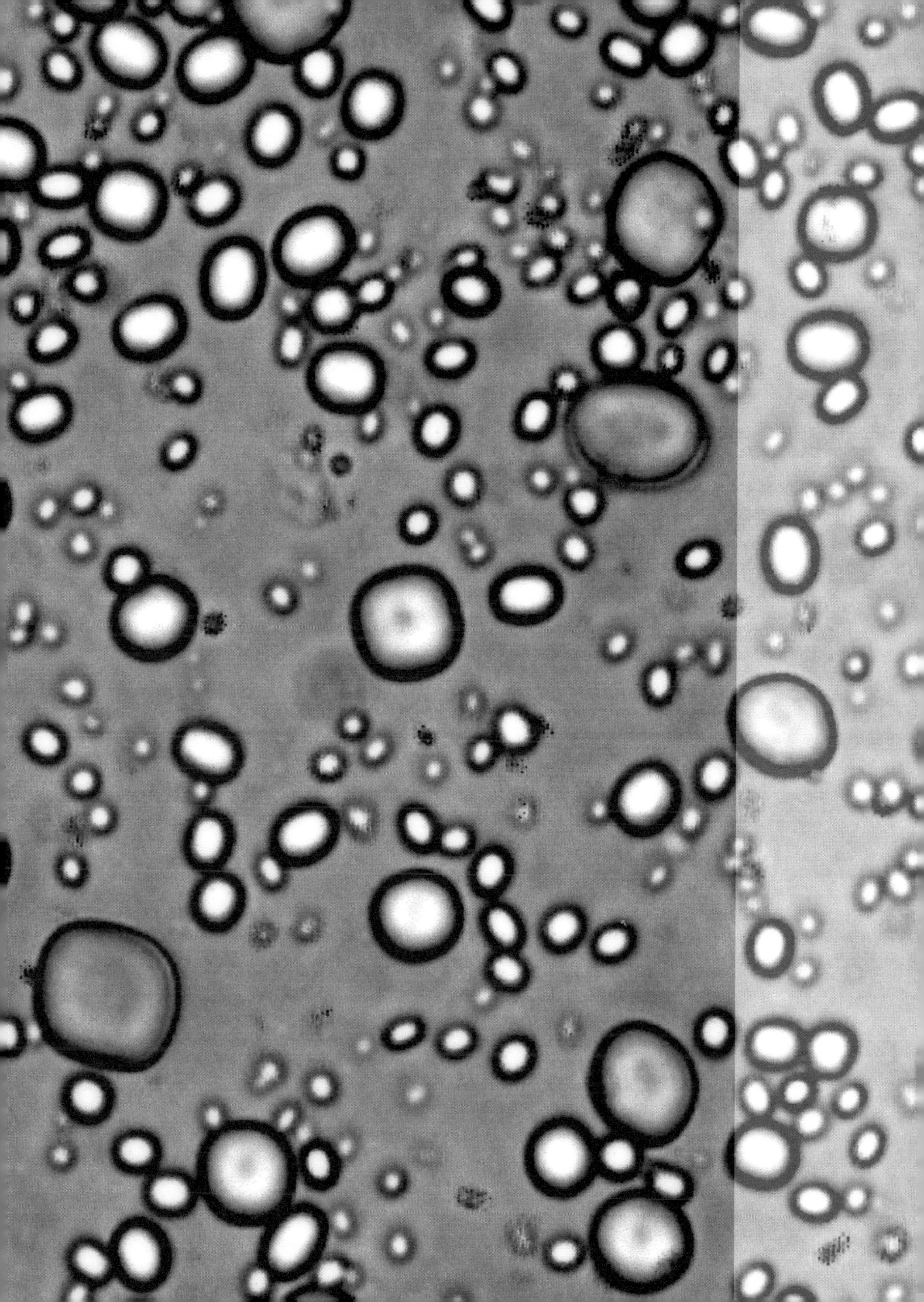

4

The Third Generation

Thomas and Yvette Karger-Weil

Heinz Karger died on Good Friday while most of Switzerland was on a holiday that lasted until the following Tuesday. That circumstance gave his 29-year-old son Thomas a little breathing space in which to cope with his grief and to prepare for the deluge of responsibilities that he had so suddenly inherited. Heinz Karger had suffered from heart disease in his last years. His death was not entirely a surprise, even though its suddenness came as a shock. Thomas Karger had known for years that he would eventually take over the company. He had been preparing for that task. When going to his father in a distant part of Switzerland he had a train ride of several hours from Basel and remembers making many notes on what to do next.

Greti Schwärzel, who had been with the company since 1948 and had served part of that time as personal secretary to Heinz Karger, still remembers the morning on which Thomas Karger told the assembled staff the sad news:'I remember we were all standing on the staircase of Arnold-Böcklin-Strasse 25 when he told us. I must say it was quiet.' Then they all went sadly back to work.

Mrs. Schwärzel said the entire staff, less than two dozen men and women at that time, all had the feeling that they wanted to help their 29-year-old director adjust to his new role. Gradually he became more and more secure in his position as head of the company. 'The publishing house was his deepest interest in life. It still is', said Mrs. Schwärzel.

The year 1959 was a watershed in two respects. Heinz Karger died, breaking the last direct link with the origins of the publishing house in Berlin, and Steven Karger, the first child of Thomas and Yvette Karger, was born. Steven, who became active in the company even before he finished his formal education, represents the fourth generation of the family in medical and scientific publishing. He also represents the first generation to be born in Switzerland. He was born in January. His grandfather lived long enough to know that there would be a fourth generation of the family to carry on the work.

From 1959 onward, the changes in the publishing enterprise have been many and profound. Heinz Karger had known each of his staff members personally and talked

Third domicile of the S. Karger Company, Arnold-Böcklin-Strasse 25, Basel (1953–1980)

1955: The whole staff at a glance

to almost all of them every day. It was comfortable to do this when the total staff was small and they were all concentrated at one address in one city. A group photograph of the staff taken on the steps of the Karger offices on Arnold-Böcklin-Strasse in 1955 shows 17 people. When the company's second century dawned in 1990 the staff numbered more than 250. But S. Karger was still a family-owned publishing house and continued to make a great effort to treat their employees as a family.

Preferred version of the publisher's emblem in use from 1946 to 1976

Thomas Karger's father had realized from the start of his years in Basel that the future of his publishing house must be international. That became even more clear after the war. This meant that any international publisher must pay far greater attention than in the past to science in the United States and Canada. For decades, Heinz Karger had depended greatly on personal contacts in the scientific world to keep abreast with what was new and potentially important. In the world of the late 1940s and the 1950s that meant he must become familiar with American science and American scientists. Accordingly, from the late 1940s onward, he took many trips west. He particularly sought out young investigators who showed promise of becoming leaders in their fields. The experience of the company from the beginning had shown that each emerging new generation of scientists was also the best source of promising authors and editors. They would be the ones with the vision, the initiative and the drive.

Heinz Karger was diligent in his efforts to find out what the new postwar generation of scientists thought their realms of research needed. Many who became authors and editors of his publications when their careers were just unfolding are now among the internationally known leaders of their fields.

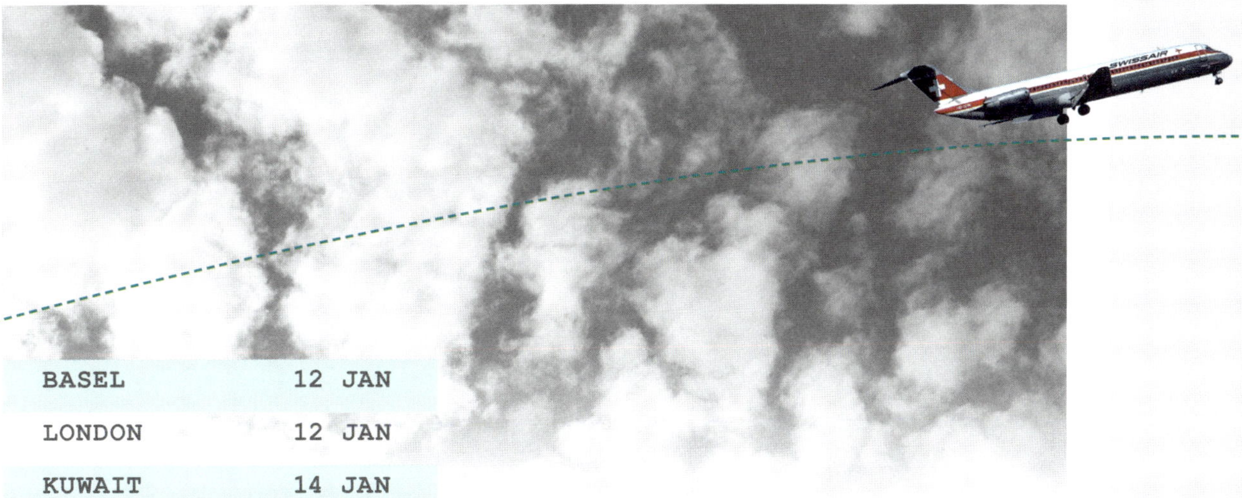

BASEL	12 JAN
LONDON	12 JAN
KUWAIT	14 JAN
BAHRAIN	15 JAN
BOMBAY	16 JAN
DELHI	18 JAN
BANGKOK	20 JAN
SINGAPORE	22 JAN
KUALA LUMPUR	23 JAN
HONG KONG	25 JAN
TAIPEI	26 JAN
SEOUL	
OSAKA	
TOKYO	
NEW YORK	
MEXICO CIT	
HAVANA	07 FEB
MADRID	08 FEB
BASEL	08 FEB

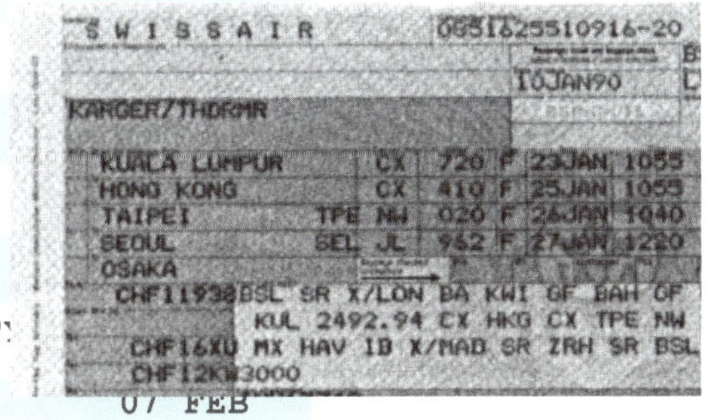

Thomas Karger has been energetic in pursuing the same policy of continual contact, personal interchange and recruitment of promising scientists in their early years. Colleagues say he travels more and at a far faster pace than his father did. He works at least as hard at the publishing business. Within a decade, Thomas Karger had established distribution centers in Paris, New York, London, Tokyo and Sydney and a publishing branch in Munich. Of course, none of this would have been achieved, Thomas Karger repeatedly said, without his wife Yvette's understanding for his devotion to his work. Yvette was often alone with their four children, assumed full responsibility for their education and took all the necessary initiatives. Thomas Karger fully realizes how deeply he, the children and the company are indebted to her.

‘In order to keep up the high standard of the Karger scientific publications … I had no choice but to switch to English as the publishing language.’

Thomas Karger (1930)

Karger Libri, Petersgraben 31, Basel, in the vicinity of the University

Karger Libri
1960
1990

Karger Libri

Thomas Karger's reputation grew with the continued success and expansion of his company. In 1960, he founded a new company: Karger Libri; with this scientific bookshop and journal agency, the Karger Group also set foot in bookselling. In a nice house in the historical center of Basel, close to the University, 45 employees – including the owner's second son, Philip – serve an international clientele.

In 1972, Thomas Karger was awarded an honorary doctor of medicine degree from the University of Hamburg, one of the institutions that had been forced by the Nazis to break all ties with the publishing house in Basel.

The publisher traveled well beyond the main centers of Europe and America. He visited many Asian countries, including Burma and mainland China, the latter in 1975, a time when such visits from the Western world were not at all common. His insistence on the international scope of S. Karger is reflected in the company's regular presence at congresses (some 450 a year) and important book exhibitions, like the 'Frankfurter Buchmesse' (Francfort Book Fair), the 'Leipziger Frühjahrsmesse' (Leipzig Spring Fair), the London International Book Fair and the exhibition of the Federation of American Societies for Experimental Biology in the United States, where the Karger Group has its own booths, or the Beijing New Foreign Textbook Exhibition. It is also evident from its publication catalogues. Authors and editors are listed from all scientifically active regions of the world. Just one example, among many, illustrates the scope: *Neuroendocrinology*, International Journal for Basic and Clinical Studies on Neuroendocrine Relationships

[29]. The journal was founded in 1965; its editorial office is in Dallas, Tex.; its editorial board includes scientists from New York, Buenos Aires, Paris, Tokyo, Budapest and Melbourne; the authors are from all regions of the world. *Neuroendocrinology* is also an example of long and successful publishing in neurosciences, one of the fields on which the company's publishing program lays most emphasis.

It was also Thomas Karger's decision to make English the prime language for the Karger publications. In this he was both continuing and advancing the policies his father had followed ever since the company left Berlin: 'My father quickly recognized that if he was to strengthen the significance of the company he would have to work for the whole European and even international scientific community', Thomas Karger noted in a speech a few years ago. 'He was one of the first medical publishers to accept German, French and English contributions for the journals and to offer his readers summaries in all three languages'.

Archiv

für

Verdauungs-Krankheiten

Gastroenterologia

DIGESTION

From the start of the Thomas Karger era, the emphasis shifted abruptly from 'European and even international' to 'international and particularly American': 'In order to keep up the high standard of the Karger scientific publications and not to lose the connection with the leading North American (English speaking) research, I had no choice but to switch to English as the publishing language', Thomas Karger said. He noted that this also guaranteed the worldwide usefulness and distribution of the company's publications. In the 1950s and 1960s, almost all of the company's journals were given English titles. The journal which came into the world originally as *Archiv für Verdauungs-Krankheiten* and later was transformed into *Gastroenterologia* is now *Digestion* [9]. The publication, which had made an impressive name for itself as *Monatsschrift für Geburtshülfe und Gynäkologie* (founded in 1895), and later *Gynaecologia* (1946–1969) and from 1970 to 1977 as *Gynecologic Investigation*, took on a new and expanded responsibility to medical science as *Gynecologic and Obstetric Investigation* [8]. There are many others.

Thomas Karger has not abandoned his devotion to the German language. Indeed, some of the company's publications still use it. 'German is my mother tongue, the language in which I think, speak and write', he says. 'I would like this language to be cultivated.' He also notes that universities in German-speaking regions will always need German teaching media and that scientific societies in German-speaking lands will continue to produce German publications. Some of the Karger journals and book series are still published in German, notably some publications in psychology, a branch of science that retains particularly close ties to that language. But, to contribute to the progress of science in the world at large, English has become the common pathway and the acknowledged international language of scientific publishing.

In the course of his major efforts to expand the company, Thomas Karger continued his father's energetic pursuit of ties to North American science. Senior American scientists still remember both men clearly: 'They are international citizens in the finest sense of the word', said Dr. Melnick. He characterized both the father and son as proud of the quality of their work and of the contribution that work makes to science: 'Heinz Karger was, and Tom is, a very serious person in my experience. Yet they are gracious in their European style', said Dr. Melnick. He recalls many occasions on which both men were very courteous and helpful to him.

Dr. Melnick's first meeting with the Karger family was typical of their continual search for scientific excellence wherever it could be found. Heinz Karger had been con-

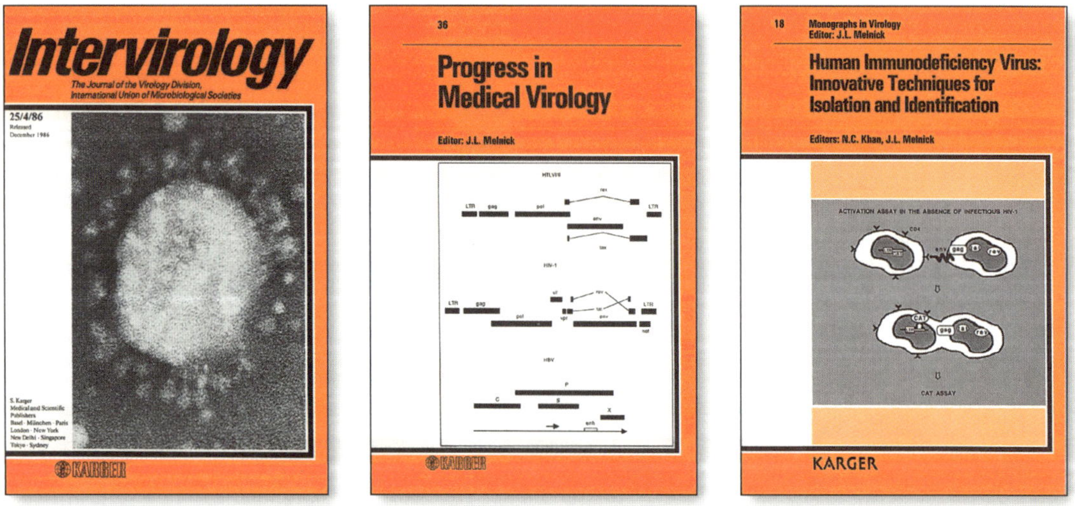

Different Karger serial publications in the field of virology:
Intervirology **– a journal,** ***Progress in Medical Virology*** **– a series,** ***Monographs in Virology*** **– a series**

Joseph L. Melnick
(1914–2001)

sulting with leaders in the field of microbiological research. They had convinced him that there was much important work to be done in virology and that Dr. Melnick was among the young investigators who were showing a lot of initiative and promise. He was already chairman, as well as the youngest member, of an important international virology panel. This was in the mid-1950s: 'He approached me, wanting to know if I would be editor of an annual on virology', Dr. Melnick recalls. The scientist is now Distinguished Services Professor of Virology and Dean of Graduate Sciences at Baylor College of Medicine.

'We came up with the title of *Progress in Medical Virology* [30] and he gave me a free hand', Dr. Melnick said. 'Thomas has given me a free hand ever since.' With Heinz Karger's encouragement, Dr. Melnick began editing the annual series of volumes to chronicle the rapid advances that were beginning to be made in studies of virus diseases. The first volume was published in 1958. In those days, medical scientists were intensely interested in the polio viruses, and the studies were beginning to turn up other viruses of the same type. Polio viruses were then causes of major epidemics in the industrialized world, but the group to which they belonged, the enteroviruses, had hardly been explored. In the mid-1950s, the only known members of this class were the three different polio viruses, a handful of less harmful agents known as coxsackie viruses for the village in upstate New York where the first one had been isolated and a few others known as echoviruses. The prefix 'echo' itself was testimony to the vast realm of the unknown that scientists were beginning to explore. 'Echo' stood for enteric cytopathogenic human orphan, a designation that gave the characteristics of the viruses with the admission that they were 'orphans' in the sense that no one had yet linked them to any diseases of mankind.

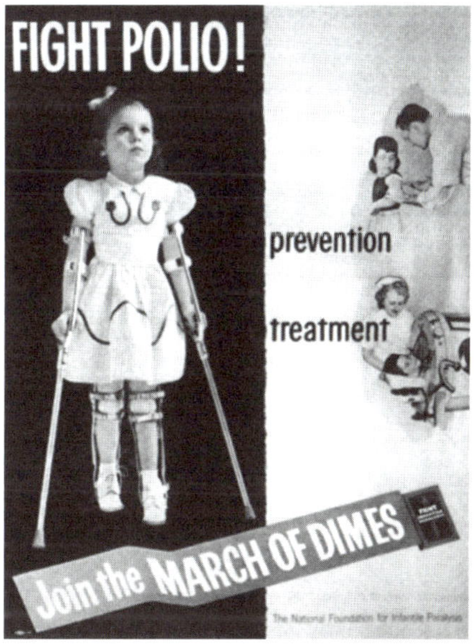

Poster used by the National Foundation for Infantile Paralyses in its 1955 campaign against polio

The growth of viruses in tissue culture was just beginning in those days and this technique brought forth a wealth of new discoveries many of which were described in Karger journals. Today, Dr. Melnick notes, over 70 different enteroviruses are known. Various members of the class have been linked to a wide range of different diseases and their molecular biology has been exhaustively studied. The nucleic acid analysis is often complete down to the identity of every base pair in the viral genetic material.

The first volume of *Progress in Medical Virology* [30] itself contained a chapter on nucleic acid as a carrier of viral activity. 'It was phenomenal at that time that we could rip off the coat and find the activity of a component of the virus particle – the nucleic acid', Dr. Melnick said. Today, he noted, there are discussions of recombinant vaccines and experimental approaches to the design of antivirus drugs. Both lines of research require detailed knowledge of every viral gene; an intensive degree of understanding that would have been hard to imagine in the 1950s. Many of the concepts that are basic to virology today were not even conceived of when the first issue of *Progress in Medical Virology* [30] was published in 1958. In addition to far broader and deeper understanding of viruses, the wide-ranging roles of these particles on the borderline between the living and inert are

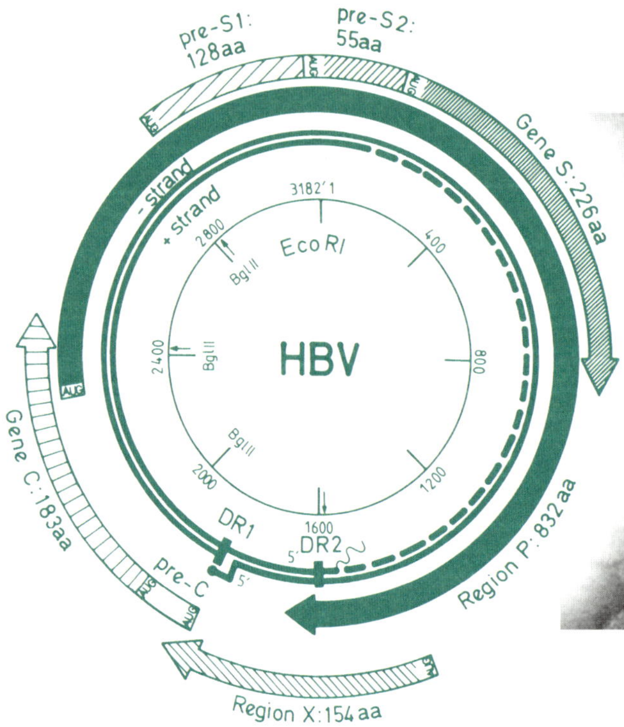

Structure and genetic organization of the hepatitis B virus (HBV) genome

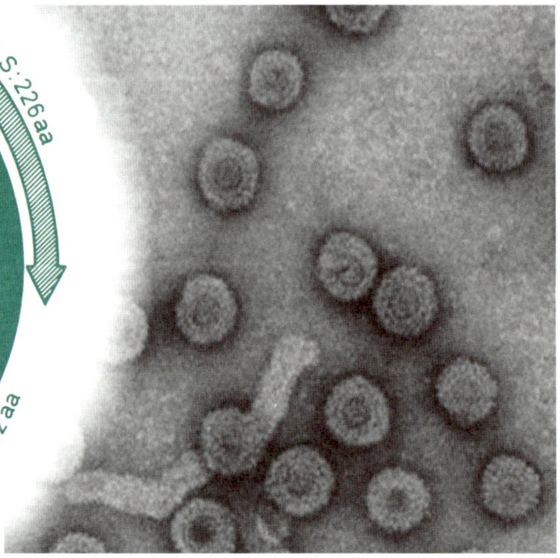

far better understood. Volume 27 of the series [31], published in 1981, contained a comprehensive set of reports on the links between hepatitis B virus and liver cancer – primary hepatocellular carcinoma. When the virology annual began, that virus was still undiscovered and any role of a virus in liver cancer would have been derided as fantasy. At present, Dr. Melnick is working on volumes 37 and 38 of the annual.

A reviewer in the *Journal of the American Medical Association* praised a recent volume of the annual as containing '... an unusual wealth of up-to-date information in a well-written form' with articles of a quality 'so high that it is well worth the attention of all'.

Another Karger virology publication, the journal *Intervirology* [32], was founded in 1973 by Dr. Melnick. This journal covers a broad range of subjects related to viruses. It has served as an important vehicle for the virology division of the International Union of Microbiological Societies in organizing scientific understanding of viruses and in bringing order and system to the complexities of viral taxonomy.

Last year (1989) *Intervirology* introduced a new section on human and related retroviruses, edited by Dr. Robert C. Gallo of the National Cancer Institute in the United States. Human retroviruses, notably HIV, the cause of the acquired immune deficiency syndrome (AIDS) constitute one of the most active frontiers of modern virology. Dr. Gallo was one of the main pioneers of AIDS virus research and is still a leader in that field.

Today, the company's catalogues still reflect creative imagination and astute perception of the frontiers of science. During the past three decades the company has continued to look for opportunities in fields that are at the cutting edge of research. Journals and book series have been inaugurated that anticipate important clinical needs. S. Karger publishes about 120 books a year on the average. In 1989 there were 85 book series. The catalogue of international scientific journals totalled 78 titles. Some of the publications, such as the journal *Nephron* [33], represent commitments to important realms of medicine that stretch back for many decades in the company's history. Others cover subjects the very existence of which might have astounded Heinz Karger viewing science from the perspective of the 1950s. The changes that have evolved from the world of Samuel Karger and his advisors a century ago are close to unimaginable.

In 1989 a new book series was in preparation on cytokines, a potentially important area of research and clinical application that hardly existed a decade ago. There are seven journals on immunology many of which cover fields and concepts that have emerged only within the past few decades of revolutionary advances in this branch of science. There are five Karger journals in pharmacology, including at least one – *Neuropsychobiology* [34] – dedicated to biological approaches to mental disorders, a concept for which powerful evidence has been amassed largely within the past 30 years and often despite bitter opposition from traditionalists.

Among the imaginative and important concepts that have found homes in Karger book series was a volume on the *Epidemiology of Natural Disasters* [35], published in 1984 in *Contributions to Epidemiology and Biostatistics* [36].

Another was *Human Monkeypox* [37], a 1988 volume that covers the biology and epidemiology of the pox viruses that can be transmitted to humans. There is a particular need to understand those viral infections now that smallpox has been erased from the earth. The volume was one in the series of *Monographs in Virology* [38], edited by Dr. Melnick.

Over a decade ago, the company began publishing the proceedings of the International Human Gene Mapping Workshops [39], in many of which Dr. Klinger was editor and some of which have been chaired by Dr. Frank H. Ruddle of Yale. The workshops, at first attended by a scant 100 or so scientists, have become increasingly important as more and more of the human archive of perhaps 100,000 genes have been mapped, assigned known functions and, in the more recent years even sequenced.

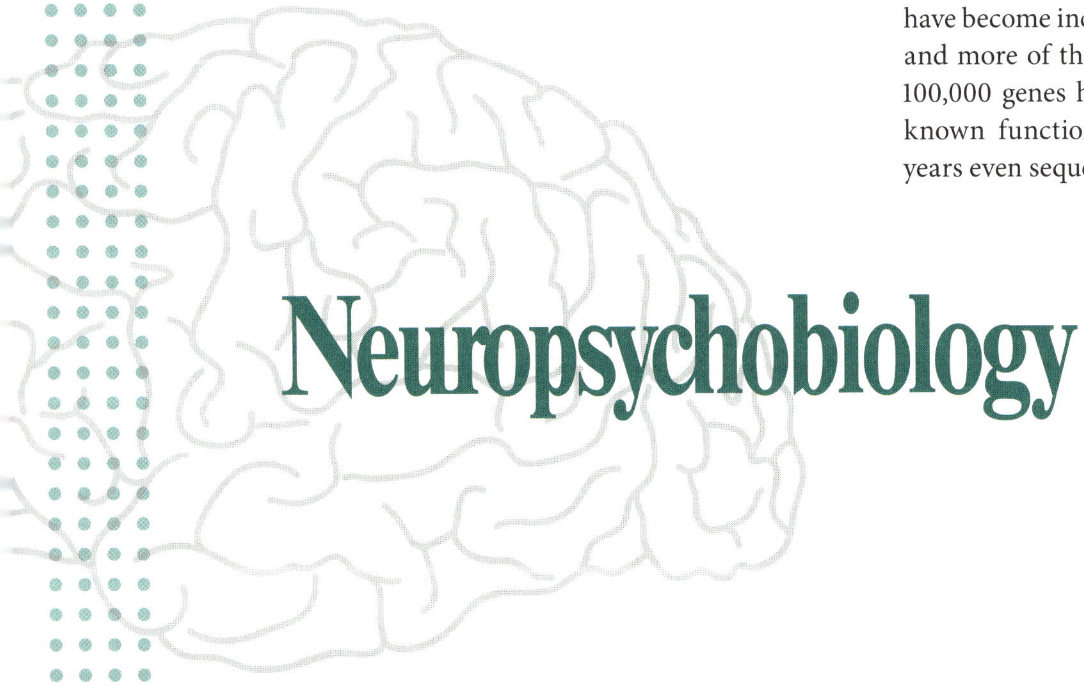

The company's publications have also broadened to include subjects far afield from medicine, but important to biologists in various specialties. S. Karger publishes *Folia Primatologica*, International Journal of Primatology [40]; the book series *Comparative Animal Nutrition* [41] that, in recent years, has covered subjects as diverse as *The Desert Camel* [42] and the *Use of Animal Models for Research in Human Nutrition* [43]. A new book series is *Comparative Physiology* [44], edited by scientists in Dortmund, FRG, and Ithaca, N.Y., USA. The series has set itself the important and intriguing task of gauging the extent to which common principles are involved in the management of cell function across species lines.

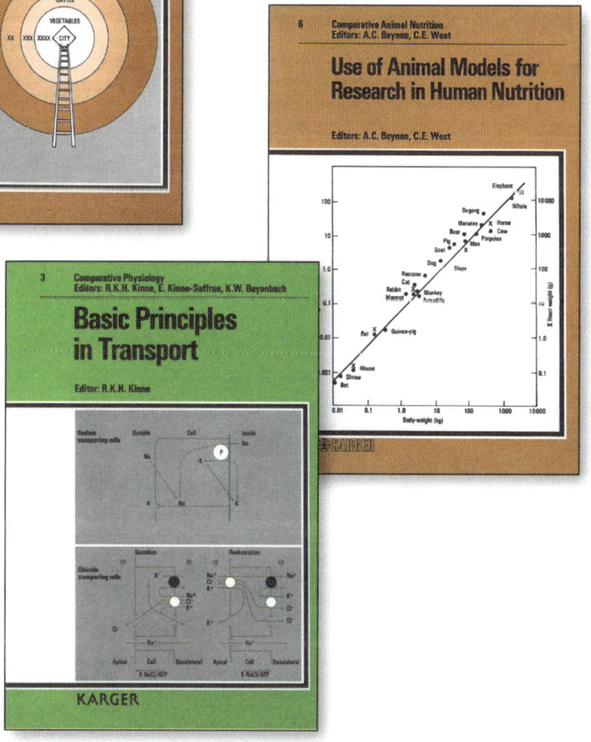

Karger Gazette

NR. 1. 1. DEZEMBER 1961

Eine neue Zeitung! Wozu? Wird nicht schon genügend Makulatur gedruckt? Auch wir haben uns diese Fragen vorgelegt, und wir verstehen Ihr Bedenken durchaus: Wann sollen Sie diese Zeitung lesen, wo Sie kaum Zeit finden, auch nur die wichtigsten Neuerscheinungen auf Ihrem engsten Interessengebiet durchzublättern? – Und dennoch: Sollten wir uns nicht bemühen, wenigstens ab und zu einen Blick über den Zaun in den Garten einer benachbarten Wissenschaft zu werfen? Sollten wir uns nicht die Zeit nehmen, uns gelegentlich mit einer uns bisher wenig vertrauten Materie auseinanderzusetzen? Unsere Zeitung möchte Ihnen hierbei behilflich sein und Ihnen mit Auszügen und Textproben aus unsern Büchern und Zeitschriften entsprechende Anregungen vermitteln. Dem eiligen Leser aber will unsere *Karger-Gazette* in Form einer Tageszeitung mit Schlagzeilen, Bildern und Anzeigen wenigstens einige flüchtige Hinweise auf unsere Verlagsproduktion geben.

Un nouveau journal!... Mais pourquoi? N'imprime-t-on pas déjà suffisamment de maculature? Nous nous sommes, nous aussi, posé ces questions et nous comprenons parfaitement vos doutes: Quand aurez-vous le loisir de lire ce journal, alors que vous trouvez à peine le temps de parcourir les nouvelles publications de votre ressort? – Et cependant: Ne devrions-nous pas, de temps en temps, jeter un coup d'œil sur d'autres domaines scientifiques et nous intéresser à des problèmes qui nous sont jusqu'ici peu familiers? Le but de notre journal est de vous venir en aide et de vous faire des suggestions en vous présentant des extraits de nos livres et périodiques. Quant au lecteur pressé, notre *Gazette*, qui paraît sous forme d'un journal avec manchettes, illustrations et annonces, voudrait lui donner quelques indications sommaires sur les ouvrages que nous éditons.

Yet another new publication! Whatever for? Aren't there enough already? Those were our own reflections and we can fully understand your objections: how could you afford the time to read this one when you barely manage to glance through the latest publication in your own particular field of interest? And yet – should we not, at least now and again, take the trouble to find out what is happening in other branches of science? Should we not occasionally make an attempt to snatch sufficient time to get acquainted with problems which have perhaps so far not been familiar to us? Our aim is to offer interesting suggestions to assist you in this endeavour by publishing extracts and quotations from the texts of our books and journals. Edited in the form of a newspaper with headlines, illustrations and advertisements, our *Gazette* is designed to provide a quick survey of the variety of *Karger* publications.

WOLF-HEIDEGGER
ALLE 3 BÄNDE
ERSCHIENEN

ATLAS DER SYSTEMATISCHEN ANATOMIE

Almost 30 years ago, early in his tenure as director of the publishing house, Thomas Karger also established a different kind of publication. This is the *Karger Gazette* [45], similar to a tabloid newspaper in format, but dedicated to reviews and synopses of important trends. The topics, usually taken from reports and reviews that first appeared in other Karger publications, have covered many of the most exciting, and at times controversial, trends in biomedical science. In an anniversary issue in 1989, Thomas Karger promised that his company's publications would continue to report on the formation of new subspecialties, the investigation of recently discovered diseases and the eradication, through the applications of medical science, of diseases that have long afflicted mankind.

The publisher noted also that 'a new and previously unheard of aspect of medicine has emerged: medical criticism' in which doctors now face moral and ethical questions some of which never had to be confronted before. Such questions have become familiar in the 1980s and seem certain to be increasingly important in the 1990s and thereafter: Is it moral to keep a human body alive simply because machinery and technology make it possible to do so? To what extent is it permissible to rebuild man and for what purposes?

In that anniversary issue of the *Karger Gazette* [45], Thomas Karger looked forward to the beginning of his company's second century. As the incredible decade of the 1980s ended, the publisher was already setting his sights on an even more portentous and incredible era in the years ahead.

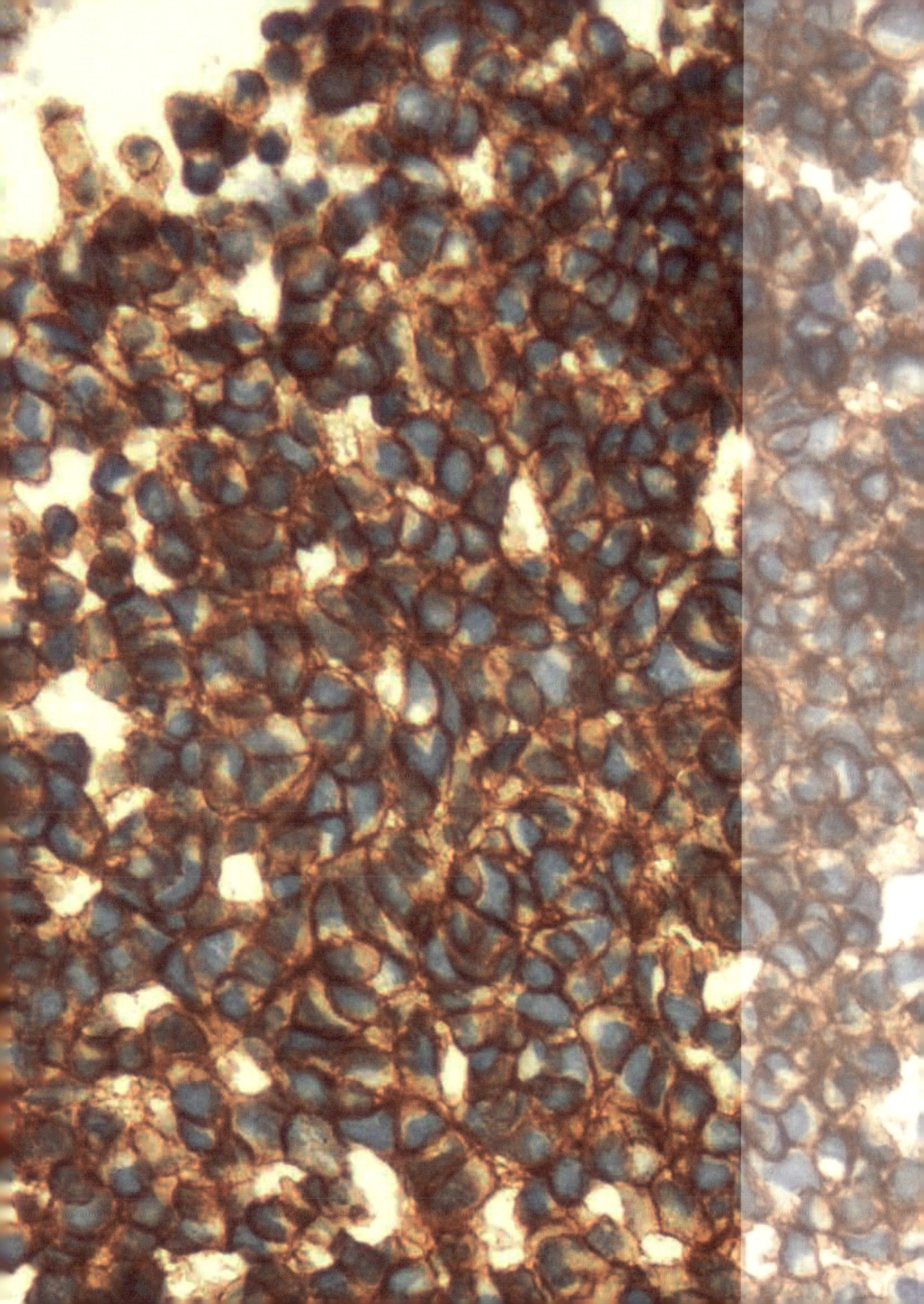

5

New Challenges

KARGER

As the decade of the 1980s closed, S. Karger faced opportunities that seemed even more exciting than Samuel Karger could have seen a century earlier. But there were also problems of a sort that the founder of the publishing house never had to face.

The opportunities were those of an era of unequalled scientific advance. Probably there has never been a previous epoch in the whole history of science and intellectual adventure to equal that of the closing decades of the 20th century. New frontiers in biology and medicine were advancing at a breathtaking pace. Virology was closing in on many of mankind's most mysterious and terrible diseases. The science of genetics, through its many subspecialties in molecular biology, was producing an entirely new and sometimes astonishingly intimate understanding of life. New clinical specialties were arising. Familiar fields of medicine were changing and expanding to an extent that would have been hardly conceivable in the past.

Publisher's emblem in use from 1977 to 1979

S. Karger has had a long history in some of the most active fields. The company's journal *Nephron* [33] is believed to be the oldest journal devoted to kidney research in the English language. Dr. G.M. Berlyne, of the US Veterans Administration Hospital in Brooklyn, N.Y., one of the publisher's senior editors in that field, said *Nephron* has published many milestone reports in scientific understanding of kidney disease. Another Karger publication, *Contributions to Nephrology* [46], carries in depth reviews and new work reported at scientific meetings, published promptly enough to keep readers up to date. Dr. Berlyne described another Karger journal, *Renal Physiology and Biochemistry* [47], as the only journal in the world devoted to research in physiology, pharmacology and anatomy focussed entirely on the kidney.

S. Karger has been publishing important works in allergy and immunology since the late 1930s. Some of the leaders in the field are editors of the company's journals and book series. In 1966, only a half dozen years after succeeding his father as publisher, Thomas Karger arranged the launching of the series, *Monographs in Allergy* [48], of which the 27th volume will be published in 1990. Karger editors believe the house is one of the world's leaders in publishing the literature of immunology.

NEPHRON

G.M. Berlyne (1931–2007)

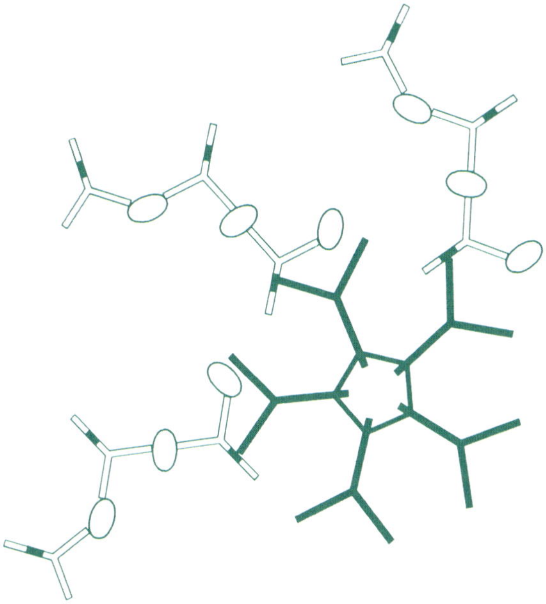

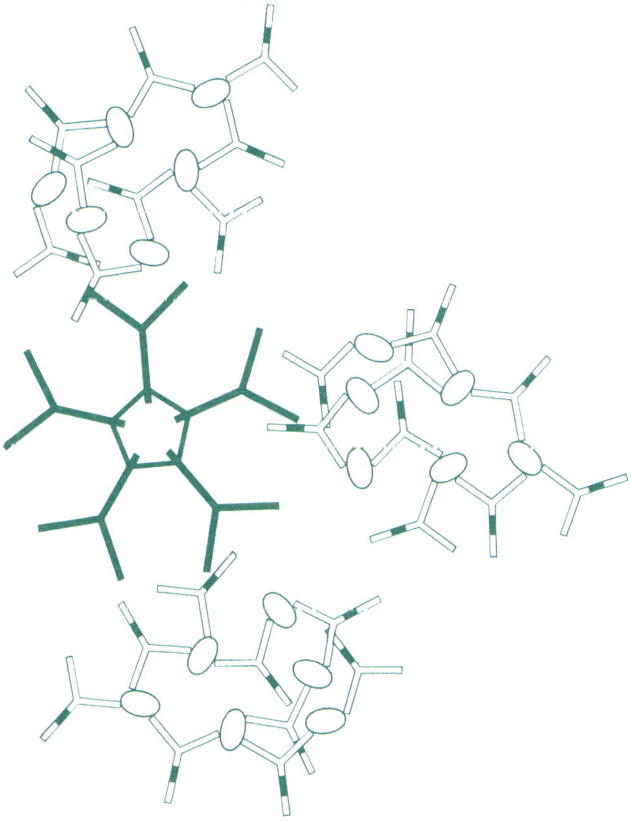

Interactions between IgG immune complexes, IgM RF and C1. After binding to soluble IgG complexes, IgM RF may remain in its nonactivating flat conformation

S. Karger was among the first to take a strong interest in AIDS, publishing in 1984 a book based on the first workshop of the European study group on the epidemic. The volume was one in the publisher's series on *Antibiotics and Chemotherapy* [49]. The workshop was held in June 1983 in Naples. It was one of the first major conferences to confront the spreading AIDS problem. In the meantime, many other publications appeared on this subject, the newest being: *AIDS and Associated Cancers in Africa* [50] and *Modern Pathology of AIDS and Other Retroviral Infections* [51]. In addition to its importance as a health concern throughout the world, the deadly disease is a mystery centered on two fundamental fields of science – virology and immunology, in both of which Karger books and journals have made important contributions.

But, while the medical and biological sciences have been advancing, the economy of publishing has been changing drastically. At the same time, the field of electronics – almost as robust as that of medicine – was threatening to produce a state of havoc in the uses of the printed word. The still expanding revolution in electronics is beginning to seem almost comparable to the one introduced by Gutenberg himself more than 500 years ago. For centuries it had been true that the printer had the last word in any potent transmission of ideas. What was on the printed page was the record and the embodiment of fact and philosophy. This gave publishers immense power and opportunity. They were the custodians of the intellect and the archivers of science. The power of print also insured order and continuity in the progress of any important field. Much of that order was be-

Fourth domicile of the S. Karger Company, Allschwilerstrasse 10, Basel (since 1980)

ginning to unravel as the last decade of the 20th century opened. Computers with prodigious memories and instant recall, electronic data transmission, copying machines, word processing and photographic printing equipment were in widespread use almost everywhere. More and more of the data and literature of science were the stored souls within these new machines. Selections and copies could be extracted and printed almost anywhere and at any time. It was not so much that users were deliberately flouting copyrights as it was that the extraction was so quick and simple that the old rules simply fell apart.

At the same time, there was an international trend toward consolidation of publishing with some houses becoming parts of huge, global conglomerates at the expense of others. The number of independent publishers has dwindled. The independence cherished at S. Karger, which had moved to new, larger headquarters at Allschwilerstrasse 10 in Basel, in the spring of 1980, still has advantages. Officers of the company note that scientists and editors know the people with whom they work at Karger. It is never just an anonymous voice or an impersonal response from a computer. The publisher can make decisions promptly, often instantly, free from the need to steer a project through a large corporate bureaucracy. Individual initiative and imagination have a fertile field in which to flower. Nevertheless, the changes in world publishing mean that an independent company such as S. Karger has to compete with giants.

While so many changes were jolting the field of medical publication, the increasing costs of research and the contraction of budgets for literature were producing problems of their own. All publishers had to cope with these problems. S. Karger was no exception.

'The world has changed. It is a fact that we have to put more emphasis on sales', said Roberto Zoppi, technical director of S. Karger. In the 1960s and through the mid-1970s, he said, medical publishing was a seller's market. Since then it has been more and more a buyer's market.

Publisher's emblem in use from 1980 to 1989

The trend was already apparent in the early 1980s when Mary-Kay Kindhauser produced a major analysis of marketing for Karger books and journals: 'The proliferation of medical and scientific publications is so great that scientists now face far more books and journals that *should* be read than time allows', she wrote in the introduction to the study. 'Since they cannot read all they should, they will select only what *must* be read.'

The situation put a premium on concise scientific literature. Among other things she said the new state of affairs meant that 'a short, compact book which covers its topic adequately but succinctly is beloved'.

As to journals, Mrs. Kindhauser's analysis noted that the brunt of the problem falls most heavily on highly focussed specialty and subspecialty journals, a category which included many of the Karger publications. For such highly specialized books and journals, the need remained great, but the number of buyers was shrinking. In the new economic climate, scientists increasingly depended on their libraries for the publications rather than having their own copies. In terms of publishing economics this meant fewer sales. The trend required higher prices. Among both books and journals, those produced by S. Karger are often quite costly, a reflection of the high quality and great pains that go into the publishing and also a response to the need to make a profit on relatively small sales volume.

The situation has been aggravated by the very circumstance that sets a publisher of serious medical and scientific works apart from the rest of commerce. Such a publisher is more than just a manufacturer. The products do significant service to mankind, contributing to conquest of disease and understanding of life. For such products there can be no compromise on the quality of content and design. But this special role also exacts a price: 'Because the material we publish is in the field of medicine, we are often considered a 'service organization' by our environment', Mr. Zoppi said in a speech to a group of advertising executives in Basel. 'Undoubtedly, medicine serves mankind, and therefore all doctors, scientists, hospital administrators, WHO officials and medical publishers are cast together in the same category. This association with humanitarianism certainly is not bad, but it also causes certain problems.' One of the problems, also shared individually by most doctors and scientists, is that of economics. While contributing to progress, one must remain solvent. For a book publisher

'Since scientists cannot read all they *should*, they will select only what *must* be read.'

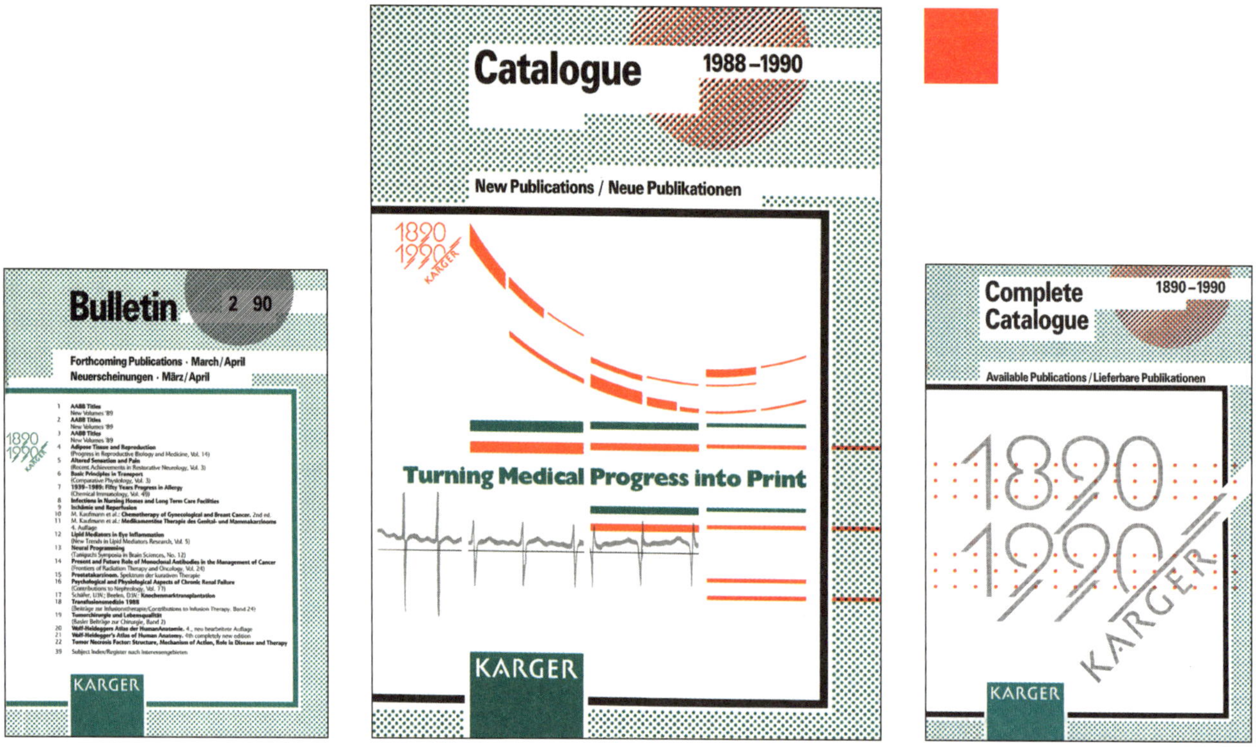

Documentation media: *Bulletin* (monthly), *Catalogue* (yearly), *Complete Catalogue* (every 3 years)

this means making at least a modest profit on the books and journals in order to survive. Like many other publishers, S. Karger has resorted to page charges to authors under some circumstances, a policy that forces scientists and their institutions to bear part of the costs of publication.

The problems of increasing costs and constricting sales exert effects not just on the publishers, the scientists and the readers of scientific literature but also on the librarians who serve as a major force in making that literature widely available. Mrs. Kindhauser's analysis quoted an executive of the huge library system of the University of California on that subject: 'Not so many years ago, there were several medical publishers whose reputations for quality were so substantial that selection of their publications was automatic. Today, the seemingly endless expansion of title lists, particularly in journal publishing, makes automatic selection by publisher increasingly difficult', the medical librarian wrote. If scientists were becoming more and more concerned with what 'must' be read, librarians had to worry about honing their lists of what 'must' be ordered.

Although S. Karger has achieved remarkable success in spite of this difficult situation, more emphasis had to be laid on new informative and creative marketing strategies aimed at the potential reader. S. Karger already has reliable documentation and advertising media. The company has made consistent efforts toward ameliorating its regularly appearing bulletins and catalogues. Thanks to extensive computerization and well-trained flexible personnel, S. Karger could keep up and perhaps even improve its reputation for handling of orders, quick dispatch, reliable invoicing system and an unbureaucratic control of payments.

The trends in publishing and the apparent constriction of markets have made everyone particularly aware of reviews, provided they appear promptly enough after a book's publication. Accordingly, S. Karger established a major book review service to keep medical and scientific librarians informed on reviewers' responses to as many new Karger publications as possible. The reprints of reviews are extensive and are sent out several times a year. There were four mailings in 1989. The Karger policy is to send out the reviews regardless of whether they are favorable or critical, an evenhanded policy that is appreciated by many librarians. Each new batch of reviews serves to underscore the broad range of scientific subjects covered by the publishing house and to make clear the response of experts in the scientific community to all aspects of these publications. One of the 1989 mailings, for example, included a brief review of the latest volume in a series, *Monographs in Clinical Cytology* [52]. This 11th volume in the series was on the *Cytopathology of Pulmonary Disease* and was written by Dr. Dorothy L. Rosenthal of Los Angeles, Calif. [53]: 'Although sputum cytology was first described over 100 years ago, it has been only in the past 20 years that the greatest advances in respiratory cytopathology have taken place', said the reviewer in the journal *Chest*. He credited this primarily to the development of the fiberoptic bronchoscope and described the Karger volume as 'an excellent addition to the expanding literature on respiratory cytopathology'. The reviewer said he found the book 'well worth the $89.50 investment as an invaluable reference'.

The mailing also included a less favorable review of the same publication in *Human Pathology*. The reviewer said the monograph 'continues the good quality of this series on clinical cytology', but considered that the writing showed a tendency to use cliches and that some of the illustrations were excessively dark.

A review, in the *British Journal of Dermatology*, of another Karger book was also highly complimentary to the volume, but less forgiving on the price of $ 212.75. The book was a 544-page work on the chemistry, physiology and mechanisms of action of the pigmentary hormones, the melanotropins [54]. The reviewer noted that there have only been a few books on the subject and that the new volume was 'most timely'. The reviewer said it should be a good introduction to the subject and should have even greater appeal to established researchers in the field, but ended with an

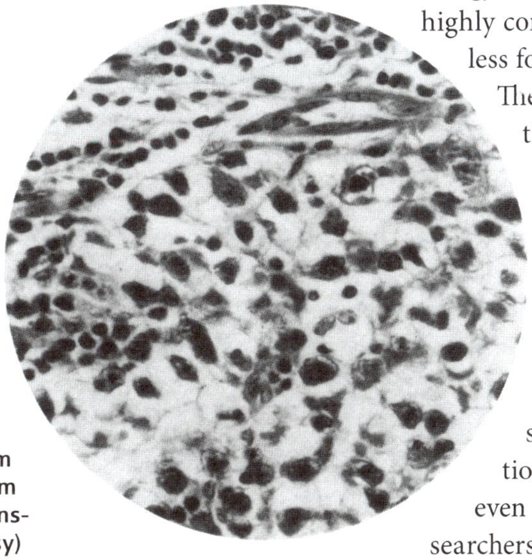

Seminoma, from the mediastinum to the lung (transbronchial biopsy)

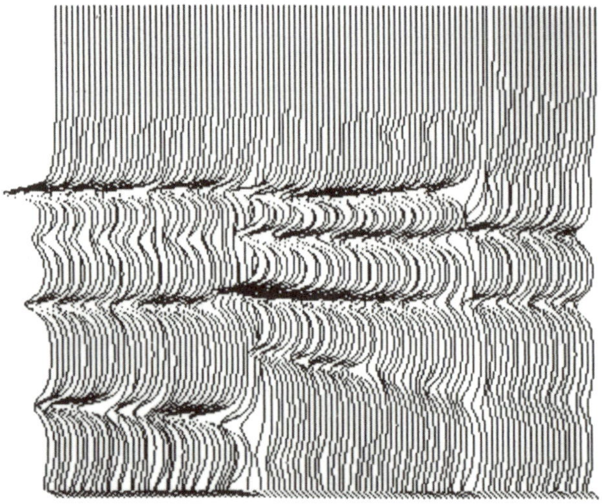

Running spectra of the vowel /u/, excised from the syllables [bu], [du], and [gu]

astringent comment on the price: 'Whether they will be able to afford the book is another matter!'

While the officers of S. Karger no doubt wince at such remarks, the reviews also help them keep abreast of what the world of science thinks of their choices of topics, the authors and the presentations. The reviews also underscore the truly international nature of modern science and the scope and wide variety of the publisher's contributions.

For example, a review, in the *South African Medical Journal*, of the 35th volume of *Progress in Medical Virology* [55] described Dr. Melnick, the founding editor of the series, as being 'one of the best known and internationally respected medical virologists' and commented that the annual volumes 'have become indispensable reading' for specialists in that field.

A review of the 24th volume of Karger's *Medicine and Sport Science* [56, 57] credited it with being 'a unique collection of studies'. The volume was on women in sports. Reviewing it in the *Journal of the American Medical Association*, Dr. Christine E. Haycock, of the University of Medicine and Dentistry of New Jersey, was particularly struck by the historical material presented, which went back as far as the athletic contests in Sparta in the heyday of Ancient Greece.

In contrast to the historical contributions of that volume, *Bibliotheca Phonetica* [58], another Karger book series, was praised for contributions to the forefront of science in an only recently emerging field. A reviewer in the technical journal *Signal Processing* cites the 12th volume of the series, *Speech and Speaker Recognition* [59], as a valuable 'working tool for a researcher or designer in the field'.

The editor of that book, Dr. Manfred R. Schroeder of Göttingen, FRG, had described speech recognition as 'one of the more glamorous technological dreams for several decades'. The reviewer described the volume as 'a stimulating book, essential for the researcher', while noting that the field itself is still 'largely prospective'.

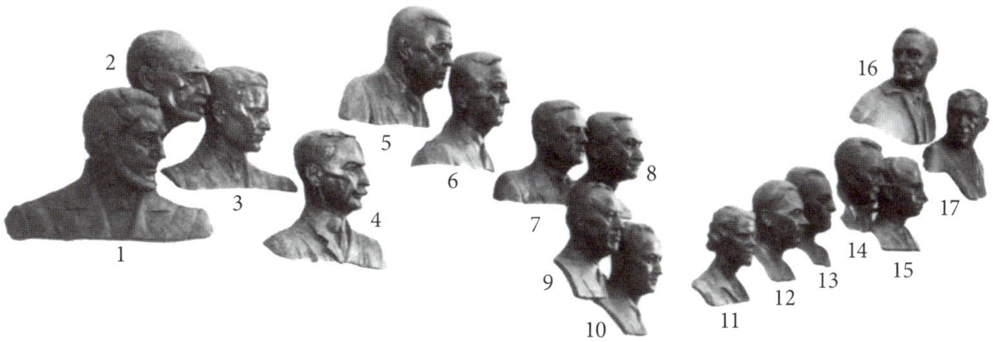

1 Jacob von Heine
2 Oskar Medin
3 Ivar Wickman
4 Karl Landsteiner
5 Thomas M. Rivers
6 Charles Armstrong
7 John R. Paul
8 Albert B. Sabin
9 Thomas Francis, Jr.
10 Joseph Melnick
11 Isabel Mørgan
12 Howard A. Howe
13 David Bodian
14 John F. Enders
15 Jonas E. Salk
16 Franklin D. Roosevelt
17 Basil O'Connor

Photograph taken in 1958 of the Polio Hall of Fame, March of Dimes headquarters, Warmsprings, Ga.

On the mundane subject of statistics, a reviewer of the Karger volume *Using and Understanding Medical Statistics*, by D.E. Matthews and V.T. Farewell of Waterloo, Ont., Canada [60], described the book as 'wonderfully clear'.

Reviews, however, can give little of the scope and duration of efforts that often go into the creation of new books. Drs. Frank Fenner and Adrian Gibbs, authors of *Portraits of Viruses* [61], published by Karger in 1988, note that their volume was first conceived a decade ago and that Dr. Melnick, then editor in chief of *Intervirology* [32], accepted their concept enthusiastically. The authors, scientists of the Australian National University, Canberra, said their idea was to ask senior virologists to prepare 'portraits' of the viruses or families of viruses that they had spent much of their lives studying.

'As in painting', the authors said, 'we wanted each portrait to reflect something of the author as well as of the subject.' The virologists themselves are an important factor

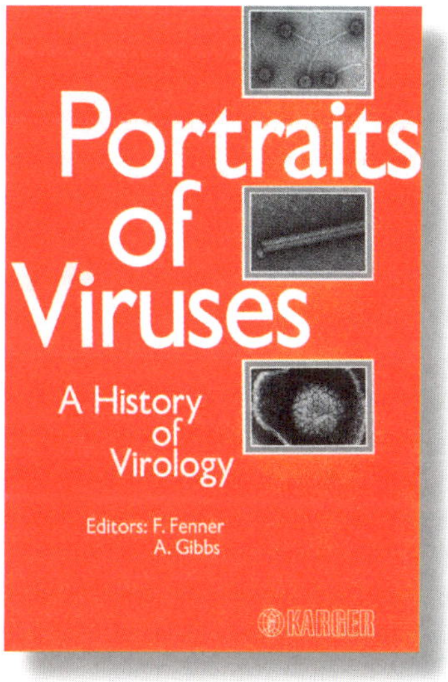

in the progress of their science. Drs. Fenner and Gibbs said, however, that they were unable to find suitable authors for the portraits of some viruses that have been critical to the development of virology. Thus the book leaves some important gaps. The authors also had hoped to include a list of 'landmark' papers in virology, but had to abandon the idea because the contributors varied so greatly in their ideas of what really constituted landmarks in the field.

As published, the collection of 15 essays covers a broad range of viruses from smallpox, known for centuries, at least through its effects, to the bacteriophages that have contributed so much to the understanding of viral infection and to the viroids, those still-mysterious virus-like agents that have only been discovered in relatively recent years. Dr. Fred Rapp, of Pennsylvania State University, the new editor in chief of *Intervirology* [32], has agreed to continue the 'Portraits' idea so that future generations of virologists can also contribute to the history of their science.

As the company has evolved, S. Karger has become more international not only in the scope of its publications, but also in its operations. Mr. Zoppi said about a third of their typesetting for books is done in Hong Kong, Ireland, and Israel. He said the quality is excellent and, given the new age of electronic communications, the production process is like dealing with 'a shop around the corner'. Most of the printing and production of books and journals is done in Switzerland and in nearby southern Germany.

KARGER

Publisher's logo in use since 1990

Being a family-run organization, S. Karger owes its personality to a large degree to that of its director, Thomas Karger. Among his colleagues and family, he has the reputation of being a hard worker who spends long hours at his business and travels far and fast in the search for new contacts, new material and new ideas. When he is at home in Basel he has everyone in his top management team into his office at least once every two days to discuss the company's operations. Associates say he produces an unending stream of new ideas which it is their task to translate into operating programs: 'He has a brilliant mind. He remembers everything and every detail', said his son, Steven. 'I would say he is very demanding, but he gives a good example. His thinking is always ahead of the present time.'

With the aid of telephone, telefax and other electronic devices, he keeps up to date on the business in Basel and his worldwide correspondence even while he is abroad.

Steven Karger has become the fourth generation of the family to enter the company. He believes that he was expected to do so from his day of birth, but that there was never any compulsion or pressure toward that end from his parents: 'Somehow, the responsibility just grew in me over a period of time while I was growing up', he said.

He entered the University of Basel to study economics in 1979, but halted his studies and went to work for the publishing house in June of 1982. His first assignment

Steven and Lea Karger-Sobol, 1990

was to work in the company's office in New York during the summer of that year. Then he returned to Basel to get a thorough, bottom to top, experience with the operations of the publishing house. Meanwhile he also continued his education in economics, with emphasis on marketing. He did this largely through correspondence courses during the next five years, finally earning a Swiss government diploma in his fields: 'During the whole period, I worked for the company', he said, noting that he preferred to combine the practical with the theoretical in that fashion, rather than simply take courses full time at a university. In recent years, he has begun to share with his father some of the world travelling and contacts with scientists that constitute the lifeblood of the enterprise. Perhaps characteristic of the family and the publishing house, his honeymoon in 1988 was a world trip that included extended business visits to all of the company's offices abroad. He and his bride, Lea, spent three weeks in New Delhi, three weeks in Tokyo and paid shorter visits to every other city in the world where S. Karger keeps a presence.

In recent years, Steven Karger has also taken part in the diplomatic interactions that are essential in dealing with a group of people as proud, independent and various as scientists. The negotiations sometimes require a well-developed sense of compromise, 'which the Swiss actually are famous for', said he, representing the first Swiss-born generation of the family.

In Basel, he has worked in every department of the publishing house to learn all aspects of the operations and to make it clear that he has no objection to hard work in any of the tasks that the enterprise involves. As of January 1, 1990, he is vice director and member of the management team. He is conscious of being its least experienced member and also of the difficulties that must sometimes arise in following in the footsteps of his dynamic and highly successful father.

Eventually, he expects to develop his own management style and foresees that it will inevitably be somewhat different from that of his father. The company is larger than it has ever been before. It is entering a new era with dramatically new technology and a world of science that is virtually exploding with new concepts and ideas. The publishing house must confront the problems of the new time and exploit its special opportunities. There is one crucial aspect of the business that he intends to keep totally unchanged. That is in maintaining close personal contacts with authors, editors, librarians and booksellers and

'The company is larger than it has ever been before. It is entering a new era with dramatically new technology and a world of science that is virtually exploding with new concepts and ideas.'

Steven Karger (1959–2008)

Liselotte Kallós-Deffner, Helsingborg

Dear Tommy, lieber Thomas,

Paul and I had originally planned to write a scientific article for this Festschrift in honour of your 60th birthday. Paul's death in October 1988 crossed this intention. Instead of this only a few lines to tell you how much the Karger-Verlag and the deep friendship between us and your parents and you meant to us over the years.

The development in Germany after 1933 and the Second World War separated us physically but your father had found a new home for your family in Basel and Paul and I went on to Sweden. The publication of the first volume of FORTSCHRITTE DER ALLERGIELEHRE – PROGRESS IN ALLERGY in 1939 – when you were ten years old – marked the beginning not only of a partnership between publisher and authors but also of a friendship. We were finally able to meet after the war in Basel and in Sweden when you and your parents came to Helsingborg for the first time in 1950. This year of your 60th birthday thus to me not only signifies more than forty years of friendship but also fifty years of uninterrupted scientific collaboration between us and the Karger family. PROGRESS IN ALLERGY was to be followed by the journal INTERNATIONAL ARCHIVES OF ALLERGY AND APPLIED IMMUNOLOGY in 1950 and later when you took over the responsibility of the Karger-Verlag our friendship as well as our scientific contacts deepened in many ways. During all these years Paul and I always felt very close to you and your family and have been happy to be able to see your children grow up and also that we could play at least a small part in the development of the Karger-Verlag.

Dear Tommy, I congratulate you with all my heart on your birthday and wish you, your family and the Karger-Verlag many successful years to come!

Yours

Liselotte Deffner-Kallós

57

Festschrift to celebrate Thomas Karger's 60th birthday

others involved in the dissemination of science throughout the world: 'What I am doing right now is just experiencing the whole business and the whole trade', he said. 'This is actually how you learn.'

Meanwhile, having brought the enterprise successfully through three decades of major expansion and change with the enthusiastic support of a dedicated and efficient management team and staff, Thomas Karger celebrates his 60th birthday in 1990.

After 1990, he hopes to relax more and do less, 'That's what he says. We will see how it comes out', said his son with a smile.

Journal publication is one of the areas of its business in which the publishing house shows its passion for efficiency most clearly. Thomas Nold, who heads the company's journal production department, says the publisher often goes to great lengths to relieve the journal editor of many of the administrative details that would raise costs and slow down the review process. For about one fifth of the 78 currently published journals, the publisher does almost everything except the selection of authors and subjects for manuscripts. Those essential tasks are the province of the editor, but all the paper work, including letters to the authors, are managed through Basel. For these journals, each editor has a secretary in Basel who handles the correspondence with authors and reviewers, registers receipt of manuscripts and their revisions and generally sees to it that the editor, wherever he or she may be, can concentrate exclusively on decision making and the creative aspects of the demanding work. Also, having a secretary in the Karger offices cements the close and personal relationship between the publisher and the editor.

Journal editorial office

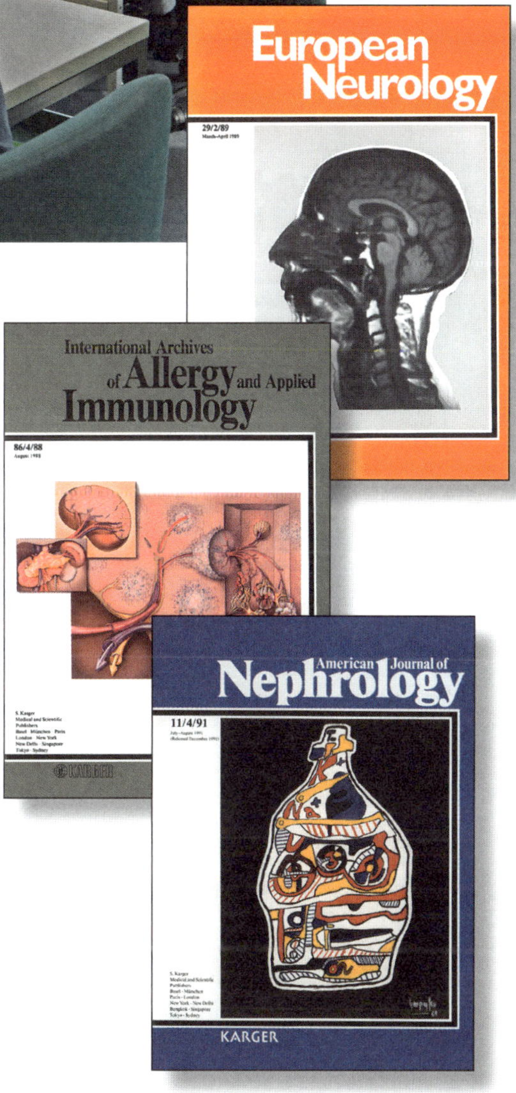

From the company's point of view, Mr. Nold says, the arrangement promotes efficiency, speed in publication and minimizes costs. One secretary in Basel can handle the affairs of several journals. The centralization of effort keeps the publishing house in close and continual touch with each journal's progress.

Throughout its worldwide operations, the publishing house continues to strive for speed and efficiency as well as excellence in publishing. The 'gestation' period of a book is about comparable to that of a human – nine months. Journals take four or five months, but can be done even more quickly when conditions are right and the need is great enough. The published proceedings from some of the international conferences of which Karger is the chronicler appear with remarkable promptness. The 10th International Human Gene Mapping Workshop [24] was held at Yale in mid June in 1989. The proceedings volume of more than 1,000 pages was produced in camera-ready form before the end of September.

But officers of the publishing house say the close association with Karger editors throughout the world is important far beyond the immediate needs of journal and book production. The editors are experts in their fields, full of ideas and strategies for bringing new developments to scientists' attention.

This is one of the main reasons why Thomas Karger and his son Steven too, spend much of their time abroad consulting with editors, learning from them, exchanging ideas and tips and, by doing all these things, preparing their company for a new decade and a new century of publishing the progress of biological science throughout the world.

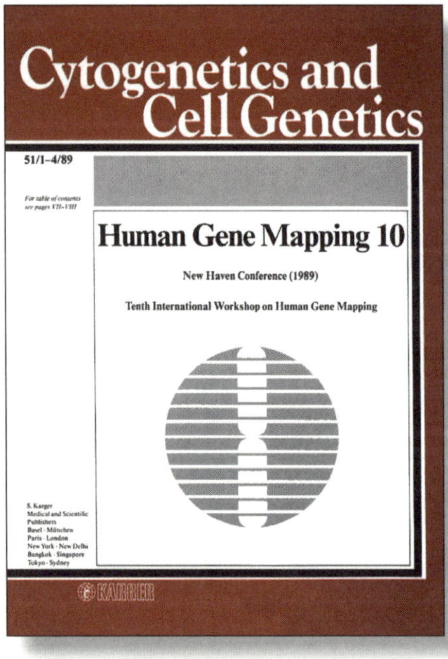

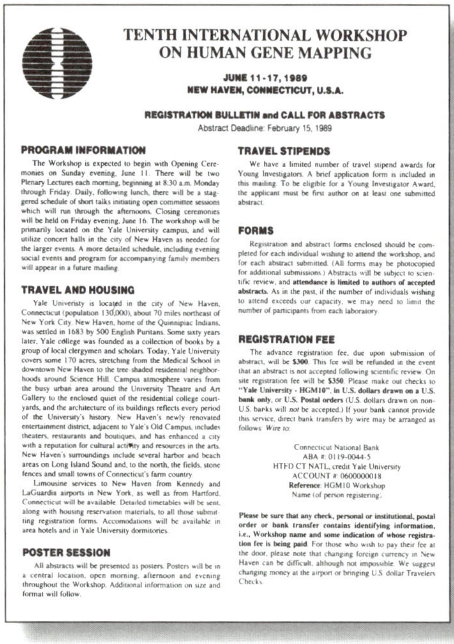

TENTH INTERNATIONAL WORKSHOP ON HUMAN GENE MAPPING

JUNE 11-17, 1989
NEW HAVEN, CONNECTICUT, U.S.A.

REGISTRATION BULLETIN and CALL FOR ABSTRACTS
Abstract Deadline: February 15, 1989

PROGRAM INFORMATION

The Workshop is expected to begin with Opening Ceremonies on Sunday evening, June 11. There will be two Plenary Lectures each morning, beginning at 8:30 a.m. Monday through Friday. Daily, following lunch, there will be a staggered schedule of short talks initiating open committee sessions which will run through the afternoons. Closing ceremonies will be held on Friday evening, June 16. The workshop will be primarily located on the Yale University campus, and will utilize concert halls in the city of New Haven as needed for the larger events. A more detailed schedule, including evening social events and program for accompanying family members will appear in a future mailing.

TRAVEL AND HOUSING

Yale Univeristy is located in the city of New Haven, Connecticut (population 130,000), about 70 miles northeast of New York City. New Haven, home of the Quinnipiac Indians, was settled in 1683 by 500 English Puritans. Some sixty years later, Yale college was founded as a collection of books by a group of local clergymen and scholars. Today, Yale University covers some 170 acres, stretching from the Medical School in downtown New Haven to the tree-shaded residential neighborhoods around Science Hill. Campus atmosphere varies from the busy urban area around the University Theatre and Art Gallery to the enclosed quiet of the residential college courtyards, and the architecture of its buildings reflects every period of the University's history. New Haven's newly renovated entertainment district, adjacent to Yale's Old Campus, includes theaters, restaurants and boutiques, and has enhanced a city with a reputation for cultural activity and resources in the arts. New Haven's surroundings include several harbor and beach areas on Long Island Sound and, to the north, the fields, stone fences and small towns of Connecticut's farm country.

Limousine services to New Haven from Kennedy and LaGuardia airports in New York, as well as from Hartford, Connecticut will be available. Detailed timetables will be sent, along with housing reservation materials, to all those submitting registration forms. Accomodations will be available in area hotels and in Yale University dormitories.

POSTER SESSION

All abstracts will be presented as posters. Posters will be in a central location, open morning, afternoon and evening throughout the Workshop. Additional information on size and format will follow.

TRAVEL STIPENDS

We have a limited number of travel stipend awards for Young Investigators. A brief application form is included in this mailing. To be eligible for a Young Investigator Award, the applicant must be first author on at least one submitted abstract.

FORMS

Registration and abstract forms enclosed should be completed for each individual wishing to attend the workshop, and for each abstract submitted. (All forms may be photocopied for additional submissions.) Abstracts will be subject to scientific review, and **attendance is limited to authors of accepted abstracts.** As in the past, if the number of individuals wishing to attend exceeds our capacity, we may need to limit the number of participants from each laboratory.

REGISTRATION FEE

The advance registration fee, due upon submission of abstract, will be **$300**. This fee will be refunded in the event that an abstract is not accepted following scientific review. On site registration fee will be **$350**. Please make out checks to **"Yale University - HGM10", in U.S. dollars drawn on a U.S. bank only**, or **U.S. Postal orders** (U.S. dollars drawn on non-U.S. banks will *not* be accepted.) If your bank cannot provide this service, direct bank transfers by wire may be arranged as follows: *Wire to:*

Connecticut National Bank
ABA #: 0119-0044-5
HTFD CT NATL, credit Yale University
ACCOUNT #: 0600000018
Reference: HGM10 Workshop
Name (of person registering)

Please be sure that any check, personal or institutional, postal order or bank transfer contains identifying information, i.e., Workshop name and some indication of whose registration fee is being paid. For those who wish to pay their fee at the door, please note that changing foreign currency in New Haven can be difficult, although not impossible. We suggest changing money at the airport or bringing U.S. dollar Travelers Checks.

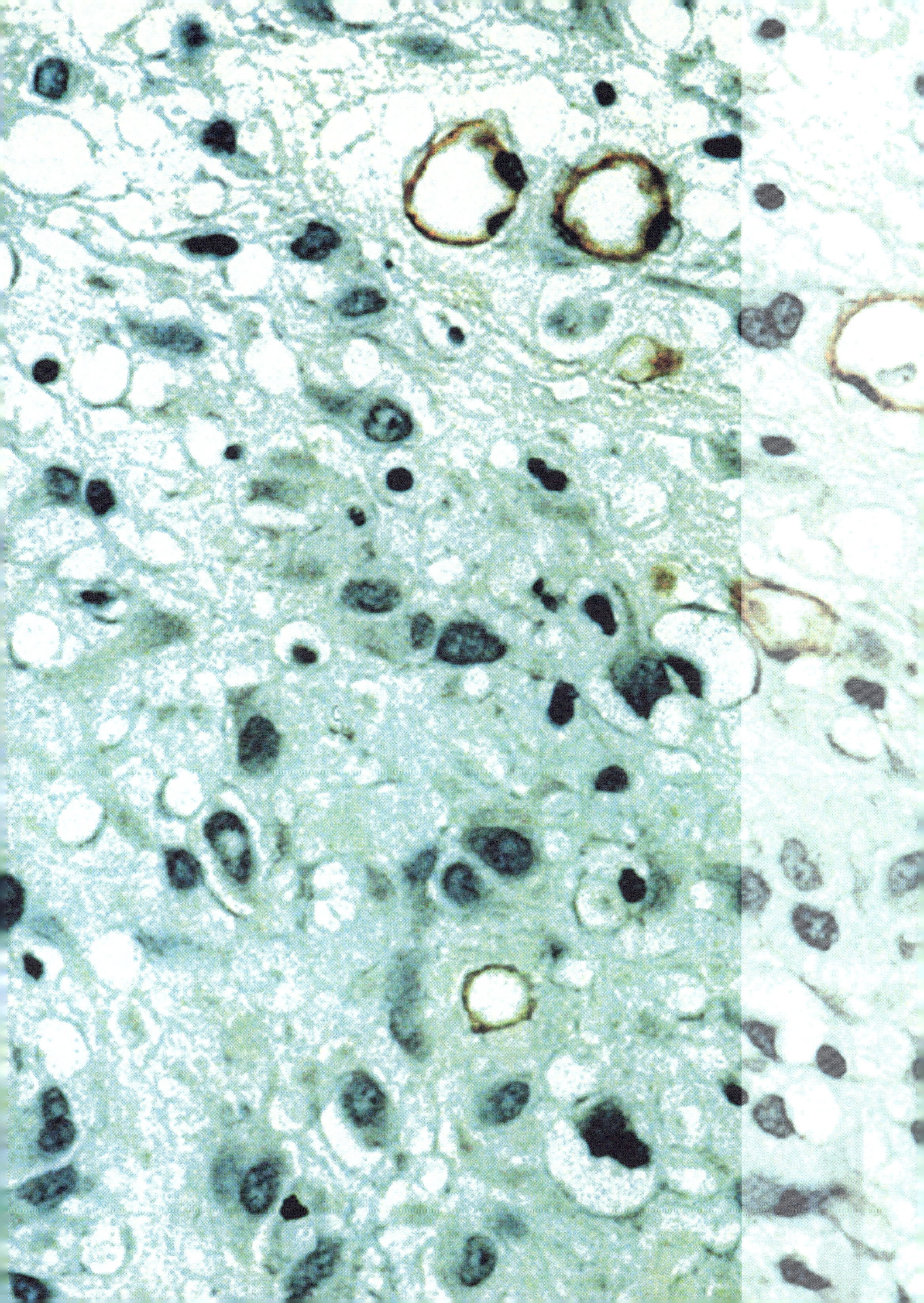

6

Entering the Second Century

1890
1990
KARGER

1990

KargerGazette

Dear Reader:

Over the period of a little less than one generation, the *Karger Gazette* has faithfully mirrored the tremendous changes that have occurred in medicine and science. For this 50th anniversary issue, I have chosen, together with our editors, contributions from the field of immunology/allergy, nephrology and cardiology, traditionally strongly represented in our publishing program. These articles are highlighting ongoing research where, as in many other promising areas, we are most likely to see innovative developments in the future.

From its inception, the *Karger Gazette* was more than a promotional tool designed to acquaint its readers with Karger Publishing's latest products and projects. Its first 49 issues are full of the most up-to-date medical findings, thanks to the world-renowned scientists who graciously made the results of their research available to this publication.

Even a simple listing of the innovations and developments which have occurred in the 28 years of the *Karger Gazette*'s existence would exceed the scope of this brief review. There is practically no area of human medicine in which spectacular progress has not been made. Once considered rare and/or expensive, today such modern technology as imaging procedures, critical care medicine, revolutionary new methods of laboratory analysis, and use of computers are all standard components of medical research. Current medicine is capable of producing results which border on the miraculous, and the *Karger Gazette* will continue to report on the formation of new subspecialities, the investigation of recently discovered disease entities, the eradication of traditional diseases due to the development of increasingly effective vaccines, and other scientific breakthroughs as they occur.

During this same period a new and previously unheard of aspect of medicine has emerged: medical criticism. As scientific knowledge progresses, skepticism grows proportionally. Doctors are now confronted with moral and ethical questions which have never before been asked.

Should science be allowed to fulfill all its potential? Is it moral to keep a human body alive only because machinery and technology are capable of doing so? Is it permissible to restructure and rebuild man? Rather, is it not necessary to re-channel scientists' thinking so they will not simply strive for what is feasible, but question its desirability?

The future issues of the *Karger Gazette* will chronicle today's science as it confronts these issues. Our publication will continue to present articles not only worth reading, but worth thinking about.

As a medical and scientific publisher, S. Karger is an integral intermediary between researchers, clinicians, and the general public. We will continue to be the 'journalistic backbone' of the scientific community, presenting the full scope of its efforts and achievements. In 1990 our firm will celebrate 100 years as a specialist in medical publishing, with almost as many journals and book series as a testimony of our expertise and successful collaboration with the medical community.

I would like to take this opportunity to thank all the authors and editors who permitted us to include their previously published work in the *Karger Gazette*. Special mention should also be made of our publishing house staff who worked untiringly to answer editorial questions, solve design problems, and overcome all the hurdles which are inevitable in the production of a high-level publication such as the *Karger Gazette*.

Thomas Karger

Few companies survive long enough to see the dawning of a second century of successful operation. S. Karger does so as the decade of the 1990s opens. Remarkably, the publishing house is still owned and operated by the family that started the business 100 years ago. With that milestone at hand, the company is not looking back, but forward both in its operating style and the subject matter of its books and journals.

In an anniversary issue of the *Karger Gazette* [45], published in 1989, Thomas Karger addressed the future and the message to which his company has been dedicated from its beginning: 'There is practically no area of human medicine in which spectacular progress has not been made', he wrote. '...current medicine is capable of producing results which border on the miraculous, and the *Karger Gazette* will continue to report on the formation of new subspecialties, the investigation of recently discovered disease entities, the eradication of traditional diseases due to the development of increasingly effective vaccines, and other scientific breakthroughs as they occur.'

What the publisher said of the *Karger Gazette* also applies with equal or greater force to the entire output of the publishing company: 'As a medical and scientific publisher, S. Karger is an integral intermediary between researchers, clinicians, and the general public', Thomas Karger said. 'We will continue to be the "journalistic backbone" of the scientific community, presenting the full scope of its efforts and achievements.'

In fact, the company has taken some important steps beyond publishing. In 1962, Thomas Karger established the Heinz Karger Memorial Foundation in honor of his father and made it the vehicle for awarding an annual prize for excellence in biomedical research. With prizes that have increased in amount over the years, the foundation has recognized contributions in some of the most advanced and important realms of medical science. Recent awards have gone to scientists for studies of oncogenes and research on the diagnostic uses of monoclonal antibodies. The award for 1990, marking the 100th year of the publishing house, will carry a prize of 40,000 Swiss Francs.

Another major feature of the company's centenary celebration will be a scientific conference to be held at the University of Basel in September. The subject will be cell to cell interaction, an area of research that is widely believed to be among the most vital and potentially important fields being explored today. New studies on the details of interactions between cells are producing knowledge that has implications for many medical disciplines. Internationally known scientists in the field have been invited to speak at the symposium.

'We will continue to be the "journalistic backbone" of the scientific community, presenting the full scope of its efforts and achievements.'

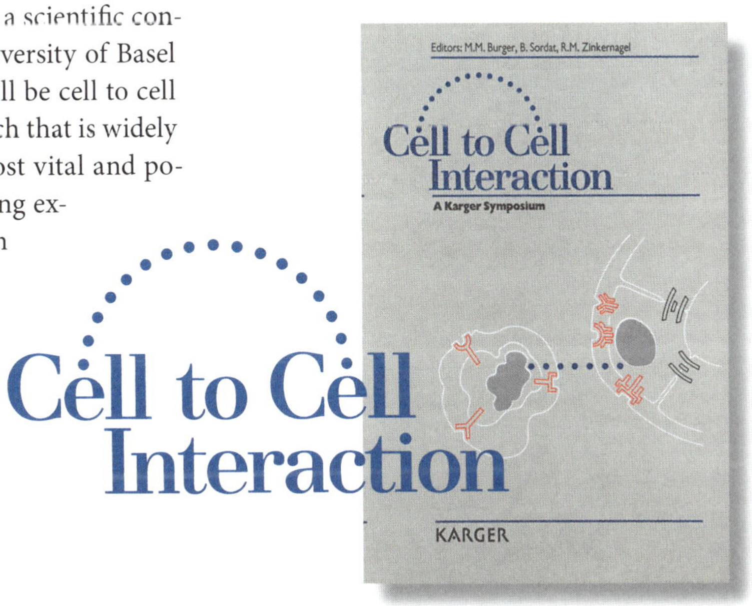

In its centennial year, the publishing house is continuing to expand its lists and to pay particular attention to subjects that lie at the most exciting frontiers of science. The 1990 catalogue lists five new book series and 12 journals that were either new, expanded in form or otherwise changed to keep abreast of science. Mr. Nold said the total of 78 journals covering many fields is a major expansion from the roughly 60 that were being published when he first came to Karger about a decade ago. Each new title is an experiment.

'We would not fulfill our task if we published only journals that bring in a lot of money', said Mr. Zoppi. 'We must have such journals, of course, to be on the safe side economically, but we also like to be in the forefront of science and there is always a risk.'

Mr. Zoppi, an expert in typography and other technical aspects of publishing, has been with the company for 25 years. He sees challenges ahead in both techniques and the scientific content of the company's publications. And he and his colleagues are continually exploring ways to streamline the publishing process.

An example is the journal *Cytogenetics and Cell Genetics* [25], one of the company's most prestigious publications. It is already passing beyond the practice of printing from manuscripts that are delivered to Basel in written form. Under the editorship of Dr. Klinger, the manuscripts are now assembled on computer disks in New York and these are sent to the publishing house in Basel. Copy-editing is done on the computer screens and, once this is complete, the disks provide the direct input for the publisher's sophisticated laser phototype setting system.

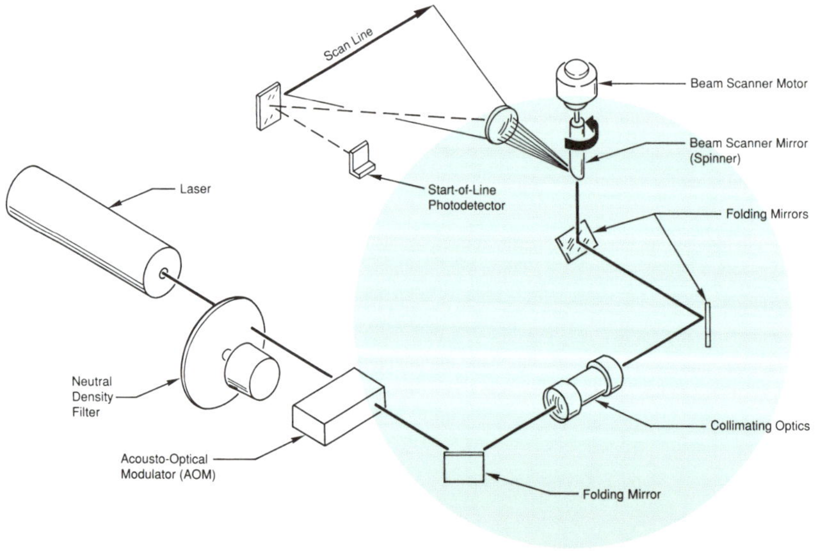

Optical path layout of a laser typesetting system

Electronic data processing in competition with…

…the traditional information media

The main advantage of putting the manuscripts on computer disks early in the publishing process is expected to be a substantial saving of time. Speed in getting into print is considered vital and the Karger technical experts think the new system may cut the journal's publishing time from the current five months per issue to three.

A similar reliance on electronic storage of some of the text material was a factor in permitting the publication within four months of the new volume of Human Gene Mapping [24], the huge volume of the proceedings of the 10th International Human Gene Mapping Workshop held at Yale in June 1989. The company expects the 11th workshop, which will be held in England in 1991, to be even more committed to the use of computer disks to mobilize, store and prepare for printing the large harvest of scientific reports that will be presented at the meeting.

The company was early in adopting computer technology and applying it to all facets of the publishing enterprise. Editorial offices and copy-editing rooms in the extension to the company headquarters are filled with the most up-to-date word processing and computer equipment. The management team of the publishing house expects the trend toward electronics to continue and to expand, and has its own data processing research and development department with a team of analysts and programmers. Yet, they do not by any means foresee the end of printed books and journals. There are still enduring and important advantages to print as repositories of the world's wisdom.

The Karger staff members admit that some of their devotion to print may be a bias of their generation. Most of the current leaders in publishing, at S. Karger and elsewhere, grew up with books and magazines – the printed page. Perhaps a new generation is arising that is equally at home before the computer screen. But print does retain some universal advantages of convenience and permanence. A well-made book will last for centuries. To use it, one needs only take it off the shelf. A person can travel more conveniently with books and journals than with

'Most fascinating is the development of new journals.'

a computer. Even at home or in one's office there are advantages to being able to mark a place in a book or even rip a particular article from a magazine or journal. And, as one of the Karger people noted, you can still read a book by candlelight when electric power fails and you can't ordinarily take a computer to the beach.

But even people deeply committed to the printed page see the electronic aids to publishing as powerfully advantageous in many respects and inevitable as part of the future of publishing. The company is already exploring the possibilities in depth.

Important as they are, all the technical aspects of publishing are simply means to an end; and this is to publish articles that advance the frontiers of medical and biological science. Mr. Zoppi notes that the much publicized philosopher of electronic communication, Marshall McLuhan, whose most famous dictum that the book was dead, long ago passed from the scene. Yet books survive and prosper, preserving the most important messages of our age between the covers of an ancient medium.

Mr. Nold is a member of the management team whose work brings him particularly close to the continual fusion of ideas and technology that brings forth each new issue of each journal. It is his department that guides the continuous stream of manuscripts through the entire sequence of editing, revision, typesetting and layout that culminate in the mailing of a new journal to its subscribers throughout the world: 'All in all it is a fascinating job', he said, 'and most fascinating is the development of new journals.' The field has changed dramatically since he first began a career in publishing. In earlier days, publishers were the prime movers in seeking out new journal ventures, he said, while now the pressure on scientists to publish their research is so great that there is hardly ever a shortage of ideas. The problem for the publisher is not in finding new subjects for journal publication, but to choose those that will be both valuable to science and useful to a large enough audience. The challenge to the publisher is to find and nurture those concepts that are most worthwhile and most likely to succeed as future trends in science develop.

Mr. Zoppi estimates that about 20 percent of the journals published by S. Karger are to some degree experimental – ideas that have come to publishing fruition, but are still awaiting the test of general acceptance by their particular scientific communities.

The *American Journal of Noninvasive Cardiology* [62], for example, will publish its fourth yearly volume in 1990, and is consciously looking towards developments in heart disease treatment and diagnosis during the first decades of the next century. *Brain Dysfunction* [63] will be in its third volume and *Dementia* [64] will be a new journal, just starting in the new decade. The journal *Neurofibromatosis* [65] began publication only two years ago, as did *Skin Pharmacology* [66], the official journal of the Skin Pharmacology Society.

Since the mainstay of Karger efforts is in highly specialized journals, the scientific communities needed for some of the new journals are quite small and select. But each such community has to account for something in the range of at least 750 subscriptions to be viable over the long run. Often, the bulk of the subscribers are libraries and research institutions. But some of the larger journals have a substantial number of individual subscribers too. A few of the company's most successful and widely read journals have subscription lists of 2,000 or more.

Karger Journals on Neurology:
- Brain, Behavior and Evolution
- Cerebrovascular Diseases
- Dementia and Geriatric Cognitive Disorders
- Developmental Neuroscience
- European Neurology
- Interventional Neurology
- Neurodegenerative Diseases
- Neuroepidemiology
- Neuropsychobiology
- Neurosignals
- Pediatric Neurosurgery
- Stereotactic and Functional Neurosurgery

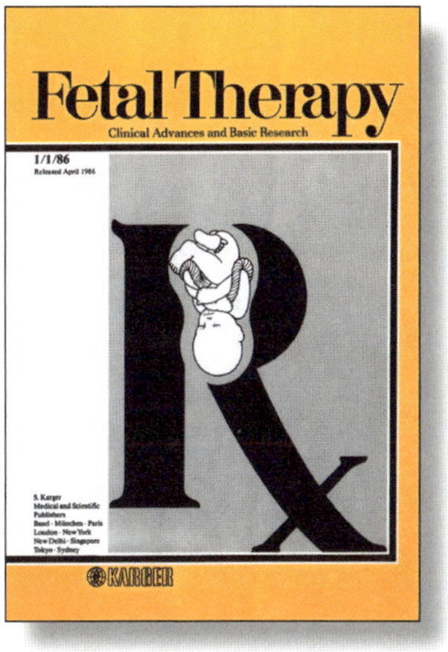

The publishing house sometimes has to nurse a new journal along for five, six or sometimes even seven years to turn the corner. Occasionally, journals have gone that distance, but then have regretfully been halted. One such case that Mr. Zoppi recalls was a journal, *Health Communications and Informatics* [67], that had several Nobel Prize winners on its board of editors. But scientists in general, it seemed, did not include it in their list of publications that *must* be read.

Mr. Nold said some of the company's scientist editors, noting the long trend toward ever greater specialization and the narrowing of research horizons, have suggested some more broadly based journals as a corrective trend. But, generally, these have not yet done well. Again, it would appear that they don't fit into the 'must read' category of enough working scientists.

Those new publications that do succeed often embody emerging fields of research and practice caught at just the right time. One such case may be the company's four-year-old journal, *Fetal Therapy* [68], even though its subscription base is still uncomfortably small. The journal first appeared in April 1986 under the editorship of two scientists in the United States, Dr. Maria Michejda, of Georgetown University Medical School in

Washington, D.C., and Dr. Kevin C. Pringle, of the University of Iowa Hospitals and Clinics in Iowa City, Iowa.

In an editorial that introduced the first issue, the editors noted that several important new techniques were making fetal diagnosis more practical and more effective than ever in the past. The same trend was developing for other aspects of fetal medicine too. For lack of a journal of their own, scientists who specialized in fetal problems were publishing in a variety of other specialized journals. The editors, and the publishing house, thought it was time to bring some unity and a fresh approach to the new field. That objective was announced in the first issue.

Dr. Michejda is now editor in chief. Dr. Pringle, now in New Zealand, Dr. W.H. Clewell, of Phoenix, Ariz., and Dr. H. Nakano of Fukuoka, Japan, are editors. In 1990, reflecting the development of the field, the journal's name was changed to *Fetal Diagnosis and Therapy* [68].

It is surely a testimony to the scientific value of the original idea that the inaugural issue also carried a guest editorial by Dr. C. Everett Koop, Surgeon General of the United States Public Health Service. Dr. Koop, one of the most active and courageous surgeons general to hold the office in modern times, was a pediatrician by training and was himself one of the founders of an important specialist publication. He congratulated the new journal on making its debut at a time when many important issues in science, medicine, philosophy and even politics were coming to focus on the field of fetal medicine.

'As I write this', he noted, 'I recall the early struggle of pediatric surgery for development and recognition for several decades before Dr. Steven Gans and I founded, and I became editor in chief of the *Journal of Pediatric Surgery*. That journal – like this one – was the first of its kind in English and had a tremendous impact on pediatric surgery here and abroad, as will this journal on fetal medicine.'

He noted that the first issue of the Karger journal would also focus on ethics 'which I urge you to keep ever before you as this field burgeons in days ahead'. One of the articles in the first issue was indeed devoted to the ethical issues of intrauterine diagnosis and therapy. Its author, Dr. LeRoy Walters, an internationally known ethicist, is director of the Center for Bioethics of the Kennedy Institute of Ethics at Georgetown University. Dr. Walters and other specialists in medical ethics were listed on the journal's editorial board.

The surgeon general's assessment of the new journal came in the first paragraph of his guest editorial: 'The new and still emerging field of fetal medicine is dignified, given a sense of purpose and direction, and expands its horizons by the appearance of this new journal', Dr. Koop wrote. 'There is no doubt that it is an idea whose time has come.'

The larger idea of a publishing house drafted on the principles of S. Karger arose in the mind of a young German bookseller in the late 19th century when the world was a far different place from what it is now. But that too was an idea whose time had come. The idea prospered in spite of adversity, catastrophe and war. Today, it is clear that the idea's time has not only come but persevered. S. Karger's grandson and great grandson are carrying that old, but still vibrant, idea forward into their company's second century.

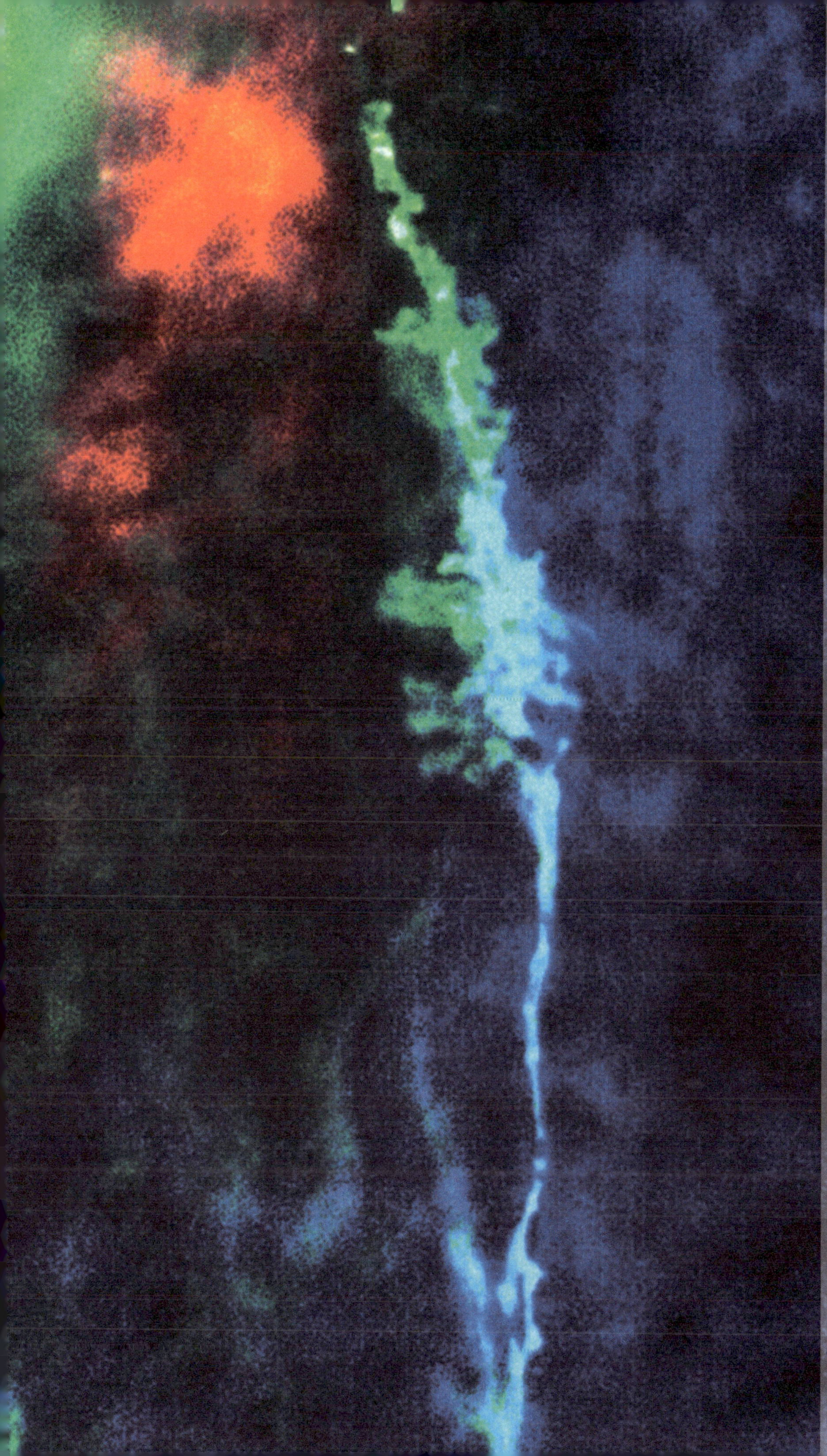

About the Author

Harold M. Schmeck, Jr., science writer for *The New York Times* for 32 years, retired as a National Science Correspondent in August 1989. He has continued to write occasional articles for *The Times* and other publications.

He joined *The Times*' staff as a science reporter in January 1957, moved to *The Times*' *Washington Bureau* as a science correspondent in 1966 and has covered the biological and physical sciences, medicine and national science policy from New York and Washington with occasional assignments abroad, including one trip to Antarctica and the South Pole in 1961.

In 1978 he returned from Washington to New York to participate in organizing *The Times*' weekly science section: Science Times.

He is author of two books: The *Semi-Artificial Man*, Walker & Company, New York 1965 (an account of the history and development of organ transplantation and use of such devices as heart pacemakers, heart valves and hemodialysis machines), and *Immunology: The Many-Edged Sword*, George Braziller, Inc., New York 1974 (an account of modern immunology research).

He authored numerous magazine articles for such publications as *Smithsonian Magazine, The New York Times Magazine, Boy's Life, Science Digest;* several articles for Encyclopaedia Britannica (medical and health annuals) and other encyclopedia yearbooks; contributed chapters to several books, and radio scripts for *Voice of America* and for *CBS News*.

Before joining the staff of *The Times* in January 1957, he worked as a reporter and science columnist for the Rochester *Times-Union* in Rochester, N.Y., from 1950 to 1957 and, before that, for a small newspaper in Illinois *(The Danville Commercial-News)*. He won numerous awards, notably several 'New York Times Publisher's Awards'.

H.M. Schmeck, Jr., was born in Tonawanda, N.Y. (a suburb of Buffalo), in 1923; he married Lois Eleanor Gallo and has one son. He studied at Cornell University (Bachelor of Arts, 1948) and at Harvard University (Nieman Fellowship, 1953–1954). From 1943 to 1945, he served as a commissioned officer in the US Army Air Forces.

Bibliography
(Chapters 1–6)

1 Dührssen A: Geburtshülfliches Vademecum. Ein kurzes Lehrbuch der Geburtshülfe für Studierende und Ärzte. Berlin, 1890. 15 editions. Translations into English, French, Italian, Polish, Russian, Turkish.

2 Dührssen A: Gynäkologisches Vademecum. Für Studierende und Ärzte. Berlin, 1890. 16 editions. Translations into English, French, Italian, Polish, Russian, Greek, Turkish.

3 Silex P: Compendium der Augenheilkunde. Für Studierende und Ärzte. Berlin, 1891. 12 editions. Translation into Italian.

4 De Ruyter G, Kirchhoff E: Compendium der allgemeinen und speziellen Chirurgie. Für Studierende und Ärzte. I. Allgemeine Chirurgie. Berlin, 1892. 12 editions. II. Spezielle Chirurgie. Berlin, 1893. 14 editions.

5 Schultz P: Compendium der Physiologie des Menschen. Berlin, 1898. 16 editions.

6 Kayser R, Klestadt W: Anleitung zur Diagnose und Therapie der Kehlkopf-, Nasen- und Ohrenkrankheiten. Berlin, 1901. 16 editions. Translation into Russian.

7 Dermatologische Zeitschrift → Dermatologica (Journal International de Dermatologie – International Journal of Dermatology – Dermatologische Zeitschrift): Founded (1893) as Dermatologische Zeitschrift by Lassar O (1893–1907); continued by Hoffmann E (1908–1938) and as Dermatologica by Lutz W (1939–1958), Schuppli R (1939–1985). Editor in chief (1990): Saurat JH, Geneva.

8 Monatsschrift für Geburtshülfe und Gynäkologie → Gynaecologia (International Monthly Review of Obstetrics and Gynecology – Revue Internationale Mensuelle d'Obstétrique et de Gynécologie – Monatsschrift für Geburtshilfe und Gynäkologie) → Gynecologic Investigation → Gynecologic and Obstétric Investigation: Founded (1895) as Monatsschrift für Geburtshülfe und Gynäkologie by Martin A, Sänger M; continued as Gynaecologia (1946–1969), Gynecologic Investigation (1970–1977) and Obstetric Investigation (1978). Former editors: Anderes E (1939–1952), Koller T (1939–1969), Käser O (1954–1969), Wenner R (1939–1967), Bloch P (1965–1969), Hermann WL (1970–1975), Keller PJ (1976–1984). Editor in chief (1990): Zador G, Södertälje.

9 Archiv für Verdauungs-Krankheiten → Gastroenterologia (International Review of Gastroenterology – Revue Internationale de Gastroentérologie) → Digestion (International Review of Gastroenterology): Founded (1895) as Archiv für Verdauungskrankheiten by Boas I; continued as Gastroenterologie (1939–1967), and Digestion (1968). Former editors: Morawitz P (1934–1936), Staehelin R (1937–1943), Hurst A (1940–1945), Löffler W (1943–1961), Hunt TC (1947–1967), Henning N (1953–1962), Ihre B (1953–1967), Bartelheimer A (1963–1967), Demole M (1963–1971), Kapp H (1968–1970), Lambert R (1972–1978). Editor in chief (1990): Creutzfeldt W, Göttingen.

10 Monatsschrift für Psychiatrie und Neurologie → Psychiatria et Neurologia (Internationale Monatsschrift für Psychiatrie und Neurologie – Revue Internationale Mensuelle de Psychiatrie et de Neurologie – International Monthly Review of Psychiatry and Neurology) → (1a) Psychiatria Clinica → (1b) Psychopathology (International Journal of Descriptive Psychopathology, Phenomenology and Clinical Diagnostics) → (2) European Neurology: Founded (1897) as Monatsschrift für Psychiatrie und Neurologie by Wernicke C, Ziehen T; continued by Bonhoeffer K (1912–1938), Klaesi J (1939–1956), Grünthal E (1953–1956), as Psychiatria et Neurologia by Klaesi J (1957–1967), Grünthal E (1957–1967), as (1a) Psychiatria Clinica by Grünthal E (1968–1973), Petrilowitsch N (1968–1970), Spoerri T (1971–1973), as (1b) Psychopathology (1984) by Berner P, Gabriel E; and (2) as European Neurology (1968) by Kaeser HE. Editors (1990) of Psychopathology: Berner P, Gabriel E, Vienna; Editor in chief (1990) of European Neurology: Kaeser HE, Basel.

11 Zeitschrift für Augenheilkunde → Ophthalmologica (Journal International d'Ophtalmologie – International Journal of Ophthalmology – Zeitschrift für Augenheilkunde): Founded (1899) as Zeitschrift für Augenheilkunde by Kuhnt H, von Michel J; continued as Ophthal-

mologica (1938/39) by Meller J (1923–1938), Behr C (1925–1938), Brückner A (1938–1959), Weve HJM (1938–1962), Dekking HM (1954–1966), ten Doesschate J (1967–1971), François J (1959–1979), Streiff EB (1954–1979), Sautter H (1979–1984). Coordinating editor (1990): Straub W, Marburg a.d. Lahn.

12 Jahrbuch für Kinderheilkunde und physische Erziehung → Annales Paediatrici (International Review of Pediatrics – Revue Internationale de Pédiatrie – Jahrbuch für Kinderheilkunde) → Pediatrie Research (An International Journal of Human Developmental Biology) → Paediatrician → Pediatrician (International Journal of Child and Adolescent Health): Founded (1858; a Karger publication since 1900) as Jahrbuch für Kinderheilkunde und physische Erziehung by Mayr F, Politzer M, Schuller M; continued by von Widerhofer H, Steffen A, Heubner O, Escherich T, von Bokay J, Feer E, Czerny A and others, as Annales Paediatrici by Freudenberg E, Bamatter F, Jaccottet M, Hottinger A, Rossi E (1938), as Pediatric Research by Lowe CU (1967–1970), as Paediatrician by Maragos GD, Greene CA (1972–1982) and as Pediatrician by Kaplan DW (1972/73). Editor in chief (1990): Kaplan DW, Denver, Colo.

13 Oppenheim H: Lehrbuch der Nervenkrankheiten. Für Ärzte u. Studierende. Berlin, 1894. 7 editions. Translations into English, Russian, Italian, Spanish.

14 Abhandlungen aus der Neurologie, Psychiatrie, Psychologie und ihren Grenzgebieten (Beihefte zur Monatsschrift für Psychiatrie und Neurologie) → Bibliotheca Psychiatrica et Neurologica (Additamenta ad Psychiatria et Neurologia) → Bibliotheca Psychiatrica: Founded (1917) as Abhandlungen aus der Neurologie, Psychiatrie, Psychologie und ihren Grenzgebieten by Bonhoeffer K (1917–1938); continued (1948) as Bibliotheca Psychiatrica et Neurologica and (1970) as Bibliotheca Psychiatrica by Klaesi J (until 1967), Grünthal E (1953–1973), Petrilowitsch N (1968–1970), Spoerri T (1971–1973). Editors (1990): Berner P, Gabriel E, Vienna.

15 Archiv für experimentelle und klinische Phonetik: Edited by Katzenstein J. Berlin, 1913/1914.

16 Phonetica (Internationale Zeitschrift für Phonetik – International Journal of Phonetics – Journal International de Phonétique): Founded (1957) and edited by Zwirner E (1957–1969); continued by Malmberg B (1962–1969), Pilch H (1962–1978), Bethge W (1963–1969), Ungeheuer G (1970–1978). Editor (1990): Kohler K, Kiel.

17 Cardiologia (Internationales Archiv für Kreislaufforschung – International Archives of Cardiology – Archives Internationales du Cœur et des Vaisseaux) → Cardiology (International Journal in Research, Prevention and Clinical Cardiology): Founded (1937) as Cardiologia by Kisch B, Löffler W; continued (1970) as Cardiology by Hegglin R (1962–1969), Mahaim I (1962–1965), Moret P (1966–1969), Galletti PM (1966–1979). Editor in chief (1990): Kellermann JJ, Tel Hashomer.

18 Acta Anatomica (International Archives of Anatomy, Histology, Embryology and Cytology – Archives Internationales d'Anatomie, d'Histologie, d'Embryologie et de Cytologie – Internationales Archiv für Anatomie, Histologie, Embryologie und Zellforschung): Founded (1945) by Chambers R, Glimstedt G, Péterfi T, Wolf-Heidegger G; continued by Wolf-Heidegger G (1945–1980), Boyden EA (1962–1974), Delmas A (1955–1980), Walberg F (1972–1980), O'Rahilly R (1981–1988). Editors in chief (1990): Lierse W, Hamburg: Goslow GE Jr, Flagstaff, Ariz.

19 Guggenheim M: Die biogenen Amine und ihre Bedeutung für die Physiologie und Pathologie des pflanzlichen und tierischen Stoffwechsels. Basel, 1940 (ed 3), 1951 (ed 4).

20 Progress in Allergy (Fortschritte der Allergielehre) → Chemical Immunology: Founded (1939) by Kallós P as Progress in Allergy (1939–1988); continued as Chemical Immunology (1989). Editors (1990): Ishizaka K, Baltimore, Md; Lachmann PJ, Cambridge; Lerner R, La Jolla, Calif; Waksman BH, New York, NY.

21 Synthetic Methods of Organic Chemistry → Theilheimer's Synthetic Methods of Organic Chemistry (An Annual Survey): Founded (1948) by Theilheimer W. Editor (1990): Finch AF, London.

22 Bericht des Ersten Internationalen Allergiekongresses – Comptes Rendus du Premier Congrès International d'Allergie – Proceedings of the First International Congress for Allergy, Zurich, 1951. Edited by Grumbach AS. Basel, 1952.

23 Primatologia (Handbuch der Primatenkunde – Handbook of Primatology – Manuel de Primatologie): Founded (1956) by Hofer H, Schultz AH, Starck D; continued by Biegert J, Kuhn HJ, Starck D, Vogel C.

24 Human Gene Mapping 10. 10th International Workshop on Human Gene Mapping, New Haven, 1989. Cytogenet Cell Genet 1989;51(1–4).

25 Cytogenetics → Cytogenetics and Cell Genetics: Founded (1962) by Klinger HP as Cytogenetics (1962–1973); continued as Cytogenetics and Cell Genetics. Editor in chief (1990): Klinger HP, Bronx, NY.

26 Wolf-Heidegger G: Atlas der systematischen Anatomie des Menschen (Atlas of Systematic Human Anatomy). Basel, 1960 (vol 1), 1961 (vol 2), 1957 (vol 3); ed 2: 1961 (vol 1), 1962 (vol 3); ed 3: 1972 (vol 1), 1971 (vol 2), 1972 (vol 3); ed 4 (1990): Wolf-Heidegger's Atlas of Human Anatomy (Wolf-Heideggers Atlas der Human-Anatomie), completely revised edition, edited by Frick H, Munich; Kummer B, Cologne; Putz R, Munich.

27 Olszewski J: The Thalamus of the *Macaca mulatta*. An Atlas for the Use with the Stereotaxic Instrument. Basel, 1952.

28 Olszewski J, Baxter D: Cytoarchitecture of the Human Brain Stem. Basel, 1954.

29 Neuroendocrinology (International Journal for Basic and Clinical Studies on Neuroendocrine Relationships): Founded (1965) by Bajusz E; continued by Knigge KM (1973–1978), Ganong WF (1979–1984). Editor in chief (1990): McCann SM, Dallas, Tex.

30 Progress in Medical Virology (Fortschritte der medizinischen Virusforschung – Progrès en Virologie Médicale): Founded (1958) by Berger E, Melnick JL. Editor (1990): Melnick JL, Houston, Tex.

31 Melnick JL, Maupas P (eds): Hepatitis B Virus and Primary Hepatocellular Carcinoma. Workshop held in Dakar, Senegal, 1980. Progr Med Virol. Basel, 1981, vol 27.

32 Intervirology: Founded (1973) by Melnick JL (1973–1985). Editor in chief (1990): Rapp F, Hershey, Pa; Editor (Human and Related Retroviruses): Gallo RC, Bethesda, Md.

33 Nephron: Founded (1964) by Richet G, Schreiner GE (1964–1971). Editors (1990): Berlyne GM, Brooklyn, NY; Giovannetti S, Pisa.

34 Neuropsychobiology (International Journal of Experimental and Clinical Research in Biological Psychiatry, Pharmacopsychiatry, Biological Psychology/Pharmacopsychology and Pharmacoelectroencephalography): Founded (1975) by Mendlewicz J; since 1983 integrating International Pharmacopsychiatry: Founded (1968) by Freyhan FA, Petrilowitsch N, Pichot P. Main editors (1990): Mendlewicz J; (Biological Psychiatry), Brussels; Pull C (Pharmacopsychiatry), Luxemburg; Janke W (Biological Psychology/Pharmacopsychology), Würzburg; Künkel H (Pharmacoelectroencephalography), Hannover.

35 Seaman J, et al: Epidemiology of Natural Disasters. Contrib Epidemiol Biostat. Basel, 1984, vol 5.

36 Contributions to Epidemiology and Biostatistics: Founded (1979) by Klingberg MA, Weatherall JAC. Editor (1990): Klingberg MA, Ness-Ziona/Tel Aviv.

37 Ježek Z, Fenner F: Human Monkeypox. Monogr Virol. Basel, 1988, vol 17.

38 Monographs in Virology: Founded (1968) by Melnick JL. Editor (1990): Melnick JL, Houston, Tex.

39 Human Gene Mapping 2. 2nd International Workshop on Human Gene Mapping, Rotterdam, 1974. Cytogenet Cell Genet 1975:14(3–6).

40 Folia Primatologica (International Journal of Primatology – Internationale Zeitschrift für Primatologie – Journal International de Primatologie): Founded (1963) by Biegert J, Hofer H, Schultz AH, Starck D; continued by Biegert J (1975–1986), Martin RD. Editor (1990): Martin RD, Zurich.

41 Comparative Animal Nutrition: Founded (1976) by Rechcigl M Jr (1976–1981). Editors (1990): Beynen AC, Utrecht; West CE, Wageningen.

42 Yagil R: The Desert Camel. Comp Anim Nutr. Basel, 1985, vol 5.

43 Beynen AC, West CE (eds): Use of Animal Models for Research in Human Nutrition. Comp Anim Nutr. Basel, 1988, vol 6.

44 Comparative Physiology: Founded (1989) by Kinne RKH, Kinne-Saffran E, Beyenbach KW. Editors (1990): Kinne RKH, Kinne-Saffran E, Dortmund; Beyenbach KW, Ithaca, NY.

45 Karger Gazette: The yearly newspaper surveying the variety of Karger publications.

46 Contributions to Nephrology: Founded (1975) by Berlyne GM, Giovannetti S. Editors (1990): Berlyne GM, Brooklyn, NY, Giovannetti S, Pisa.

47 Renal Physiology and Biochemistry (International Journal of Experimental Renal Physiology, Pathophysiology, Biochemistry and Pharmacology): Founded (1978) by Berlyne GM, Thomas S. Editors (1990): Berlyne GM, Brooklyn, NY; Lang F, Innsbruck.

48 Monographs in Allergy: Founded (1966) by Dukor P, Kallós P, Trnka Z, Waksman BH, de Weck AL. Editors (1990): Dukor P, Basel; Hanson LA, Göteborg; Kallós P†, Helsingborg; Shakib F, Derby.

49 Antibiotica et Chemotherapia (Fortschritte – Advances – Progrès) → Antibiotics and Chemotherapy: Founded (1954) as Antibiotica et Chemotherapia by Bloch H, Gsell O, Langer E; continued (1965) as Antibiotics and Chemotherapy by de Weck A, Schönfeld H (1971). Editor (1990): Schönfeld H, Grenzach.

50 Giraldo G, Beth-Giraldo E, Clumek N, Gharbi MR, Kyalwazi SK, de Thé G (eds): AIDS and Associated Cancers in Africa. 2nd International Symposium on AIDS and Associated Cancers, Naples. 1987. Basel, 1988.

51 Haase AT, Gluckman JC, Racz P (eds): Modern Pathology of AIDS and Other Retroviral Infections. Application of Contemporary Methods. International Workshop on 'In situ Hybridization of Retroviruses', Hamburg, 1988. Basel, 1990.

52 Monographs in Clinical Cytology: Founded (1965) by Wied GL. Editor (1990): Wied GL, Chicago, Ill.

53 Rosenthal DL: Cytopathology of Pulmonary Disease. Monogr Clin Cytol. Basel, 1988, vol 11.

54 Eberle AN: The Melanotropins. Chemistry, Physiology and Mechanisms of Action. Basel, 1988.

55 Melnick JL (ed): Progress in Medical Virology. Basel, 1988, vol 35.

56 Medicine and Sport Science: Founded (1969) by Jokl E (1969–1984), Hebbelinck M (1969). Editors (1990): Hebbelinck M, Brussels; Shepard RJ, Toronto, Ont.

57 Adrian MJ (ed): Sports Women. Med Sport Science. Basel, 1987, vol 24.

58 Bibliotheca Phonetica (Additamenta ad Phonetica; ref. 16): Founded (1964) by Lehiste I, Pilch H, Zwirner E; continued (1985) by Kohler K. Editor (1990): Kohler K, Kiel.

59 Schroeder MR (ed): Speech and Speaker Recognition. Bibl Phonet. Basel, 1985, vol 12.

60 Matthews DE, Farewell V: Using and Understanding Medical Statistics. Basel, 1985 (ed 1), 1988 (ed 2).

61 Fenner F, Gibbs A (eds): Portraits of Viruses. A History of Virology. Basel, 1988.

62 American Journal of Noninvasive Cardiology: Founded (1987) by Spodick DH. Editor (1990): Spodick DH, Worcester, Mass.

63 Brain Dysfunction: Founded (1988) by Sega A. Editor (1990): Sega A, Rome.

64 Dementia: Founded (1990) by Chan-Palay V. Editor in chief (1990): Chan-Palay V, Zurich.

65 Neurofibromatosis (Clinical Advances and Basic Research): Founded (1988) by Riccardi VM. Editor in chief (1990): Riccardi VM, Houston, Tex.

66 Skin Pharmacology: Founded (1988) by Schaefer H. Editor (1990): Schaefer H, Valbonne.

67 Biosciences Communications (An International Journal of Communications Research and Theory in the Biosciences and Fields of Health Care Delivery) → Health Communications and Informatics (Interdisciplinary Journal of Communications and Computer Sciences in Health Care Delivery): Founded (1975) as Biosciences Communications by Day SB; continued (1979) as Health Communications and Informatics by Day SB, New York, NY.

68 Fetal Therapy (Clinical Advances and Basic Research) → Fetal Diagnosis and Therapy (Clinical Advances and Basic Research): Founded (1986) as Fetal Therapy by Michejda M, Pringle KC; continued (1990) as Fetal Diagnosis and Therapy by Michejda M. Editor in chief (1990): Michejda M, Washington, DC.

From 'Turning Medical Progress into Print' to 'Connecting the World of Biomedical Science'

A quick comparison of the title of the 1990 Festschrift with the title of its sequel in 2015 neatly summarizes how the publishing world has evolved in the last 25 years. Not long after Karger had celebrated the publishing company's first centenary, 'print' started to lose ground – particularly in the context of scientific publications – as digital content began to increasingly supplement and then replace printed publications. Rather than looking through printed journals and books for information, it does indeed often seem to be a lot simpler, using appropriate search criteria, to quickly check through a variety of different digital sources in order to see whether anything relevant might be available. It is not my intention here to get involved in the discussion about whether the chances of making serendipitous discoveries might be reduced by the increasing trend to focus on a single topic (serendipity: the discovery purely by chance of something new, surprising or inspiring). Ultimately, the only thing that is truly important is that scientists retain their curiosity and that they agree with Isaac Asimov's statement: 'The most exciting phrase to hear in science, the one that heralds new discoveries, is not "Eureka! (I found it!)" but "That's funny...".' At the end of the day, it hardly matters whether someone arrives at an insight via printed or via digital information or whether it is the result of observation or systematic experimentation.

The world is forever changing. And it would seem it has been changing much more quickly in the last 25 years than in the first 100 years of the company's history following its founding by Samuel Karger. While revising the Festschrift published for the centenary it became obvious that, where the images evoked by Harold M. Schmeck's words are as true today as they ever were, the visual elements of the 1990 publication seem, just a quarter of a century later, to be so outdated. The solution was to give the publication a new and fresh layout. In the final section of this Festschrift we put together a selection of our current publications. The description of their content will, of course, still provide a useful guide in 25 years' time, but it is to be expected that their visual appearence will no longer attract the reader's eye in the same way they do today.

What is going to happen in the next 25 years? Will scientists who exchange information and ideas with each other still want help from a publisher like us? Kent R. Anderson, publisher of *Science* at the AAAS, has set down his thoughts on this question in the *Karger Gazette* (www.karger.com/gazette), which is also covering our 2015 anniversary. I share his perception and ideas. There would be no point to paraphrase him, and so, with his permission, we are reproducing an abridged version of his article after the last chapter in the new Festschrift (pages 165 to 171).

Writing this text in early 2015 I look forward to our celebration year. We all share the drop of bitterness that my oldest brother Steven, who passed away far too early, can't be with us. However, he would have wanted us to go on and to enjoy our festivities. So we will celebrate the 125 years of existence of the publishing house as a family-owned and family-run enterprise together with my parents, siblings and in-laws, nephews and niece, and with all our employees from all over the world who will join us in late June here in Basel.

I would like to take this opportunity to end here with my deepest thanks to all and everybody who supported Karger Publishers throughout the many years – as employees, authors, editors, partners, customers, readers or in any other role. Thank you!

Gabriella Karger

Mechthild Hempe

Karger – Connecting the World of Biomedical Science

7

Becoming a Digital Company

KARGER

In 1990, S. Karger Publishers celebrated its hundredth anniversary full of self-confidence. There was every reason to feel entrepreneurial pride as the family had guided the publishing house through serious political and economic crises for a number of generations and not only preserved but expanded it. After 100 years, Karger was successful, with a diversified program of journals and books in the field of medical research. Journals like *Dermatology*, which had first been issued in 1893 under the title *Dermatologische Zeitschrift*, also had a long history very much like the publishing house. Subsequent publications established Karger's image as a broker of scientific information.

The company's centenary celebrations began in March 1990 with an official ceremony: the publishing house invited representatives of the university, local and national government, professional associations, science and media to the Wildt'sche Haus. The Late Baroque townhouse situated on Petersplatz in the center of Basel housed such institutions as the Swiss Academy of Medical Sciences. The prestigious ambience and the academic setting suited S. Karger perfectly. Basel had long become the home base of the company that was founded in Berlin. During the ceremony, the publisher Thomas Karger once again expressed his gratitude to the city for its support during the company's move from

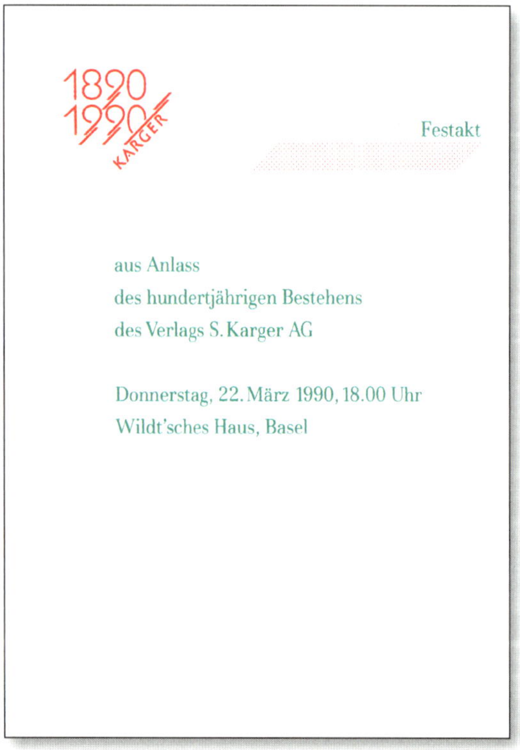
1890
1990
KARGER

Festakt

aus Anlass
des hundertjährigen Bestehens
des Verlags S. Karger AG

Donnerstag, 22. März 1990, 18.00 Uhr
Wildt'sches Haus, Basel

Programm

Kaspar Kummer (1795–1870)
Erster Satz aus dem Trio F-Dur, op. 32, für Flöte, Klarinette und Fagott
Allegro

Begrüssung durch Dr. h.c. Thomas Karger

Grussadressen
Prof. Dr. Beat Roos, Direktor des Bundesamtes für Gesundheitswesen
und Delegierter des Bundesrates, Bern
Prof. Dr. Hans-Rudolf Striebel, Regierungsrat Basel-Stadt
Peter Schmid, Regierungsrat Basel-Landschaft

Wolfgang Amadeus Mozart (1756–1791)
Erster Satz aus dem Streichquartett d-moll, KV 421
Allegro moderato

Festansprache von Dr. Peter G. Rogge, Basel
«Die Verbreitung des Wissens – Fachverlage vor neuen Herausforderungen»

Eröffnung des «Karger Fonds für Nachwuchsförderung
in Medizinischer Forschung»
Dr. h.c. Thomas Karger
Prof. Dr. Carl Rudolf Pfaltz, Rektor der Universität Basel

Wolfgang Amadeus Mozart
Oboenquartett F-Dur, KV 370
Allegro – Adagio – Rondeau: Allegro

Thomas Karger, Steven Karger and Prof. Dr. Beat Roos, Director of the Federal Office of Public Health and delegate of the Swiss Federal Goverment

Germany. He set up a trust to promote up-and-coming scientists, which continues to fund projects long after the jubilee year. Today, revenues from the assets are used for the 'Science Month' of the Medical Faculty of the University of Basel, which gives students the opportunity to attend a multi-week course to learn the fundamentals of scientific work.

In 1990, Basel was the setting for more jubilee events. In May, more than 400 employees and family members attended a gala at the Le Plaza Hotel. They enjoyed the relaxed atmosphere and personal charm of the evening. Although the staff had notably increased, the publishing house still possessed the special character of a family business, a fact which was well reflected during that evening. Employees had helped prepare the event, and Steven Karger personally took on the role of master of ceremonies on this occasion.

Remaining close to science is the primary constant for the publishing house – originating in its history and always carefully maintained. An international symposium on 'Cell to Cell Interaction' in September 1990 underlined its claim of remaining closely linked to medical research. However, whether scientific progress would always be transformed into printed work, as suggested in the company motto for the jubilee year 'Turning medical progress into print', was already open to de-

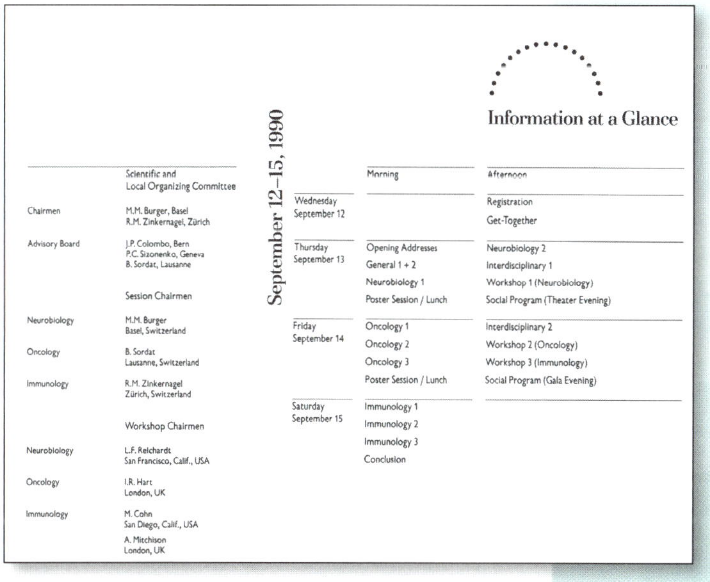

Information at a Glance

September 12–15, 1990

	Scientific and Local Organizing Committee
Chairmen	M.M. Burger, Basel R.M. Zinkernagel, Zürich
Advisory Board	J.P. Colombo, Bern P.C. Sizonenko, Geneva B. Sordat, Lausanne
	Session Chairmen
Neurobiology	M.M. Burger Basel, Switzerland
Oncology	B. Sordat Lausanne, Switzerland
Immunology	R.M. Zinkernagel Zürich, Switzerland
	Workshop Chairmen
Neurobiology	L.F. Reichardt San Francisco, Calif., USA
Oncology	I.R. Hart London, UK
Immunology	M. Cohn San Diego, Calif., USA A. Mitchison London, UK

	Morning	Afternoon
Wednesday September 12		Registration Get-Together
Thursday September 13	Opening Addresses General 1 + 2 Neurobiology 1 Poster Session / Lunch	Neurobiology 2 Interdisciplinary 1 Workshop 1 (Neurobiology) Social Program (Theater Evening)
Friday September 14	Oncology 1 Oncology 2 Oncology 3 Poster Session / Lunch	Interdisciplinary 2 Workshop 2 (Oncology) Workshop 3 (Immunology) Social Program (Gala Evening)
Saturday September 15	Immunology 1 Immunology 2 Immunology 3 Conclusion	

Program of the 'Cell to Cell Interaction' Symposium, September 12–15, 1990 in Basel

bate. Electronic media were looming on the publishing horizon. What form they would take and whether they would prevail was as unclear as how the company would calculate for its new products. Thomas Karger, who had managed the publishing house since 1959, continued to be confident. While remaining quite open to technological innovations, his chief concern was the content of the publications. Whatever the form, he aimed to continue to publish relevant information for researchers and clinicians in medicine. He expressed his wishes for the future that his publishing house would continue to grow moderately and enter the next century of the company's history under a fourth-generation family member.

The anniversary coincided with a time of change and transition. The publication of the fourth edition of Wolf-Heidegger's *Atlas of Human Anatomy* in 1990 was again a masterwork of high-quality book design and production, as was the tradition at Karger. More than ten years of work had gone into the 600-page tome – involving a re-conception, updated illustrations and numerous revisions to the text and figures, lithography and the printing process. Some of the staff of the publishing house thus worked on this classic medical publication using established production methods, while others further developed a seminal project by networking the PCs that the company had been using for many years. The vision for the future was clear: text and figure files should reach the setting phase for further processing in an uncomplicated way, and everyone involved in the production process across all departments should gain an overview of the data and how to access it.

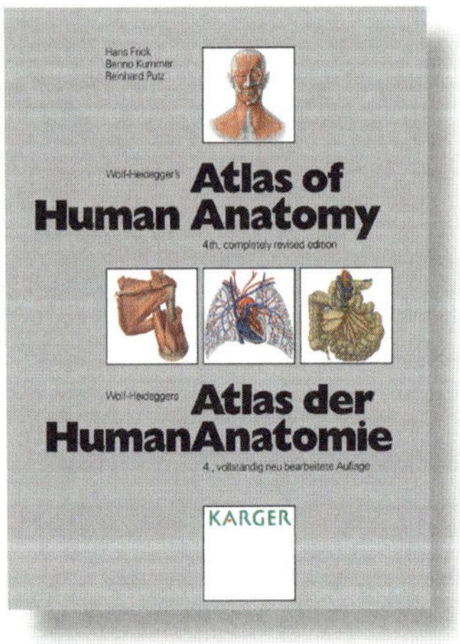

4th Edition 1990

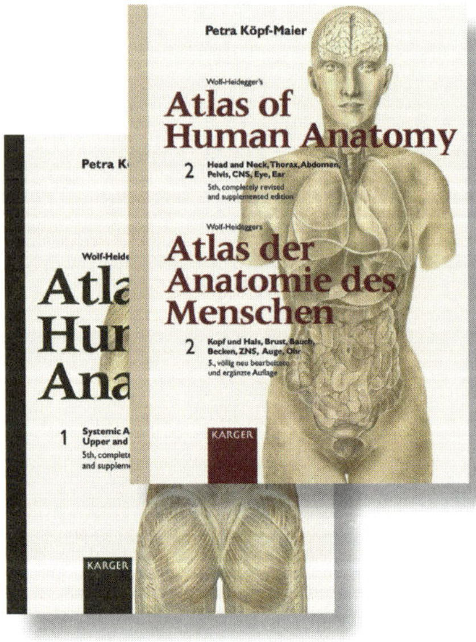

5th Edition 2001

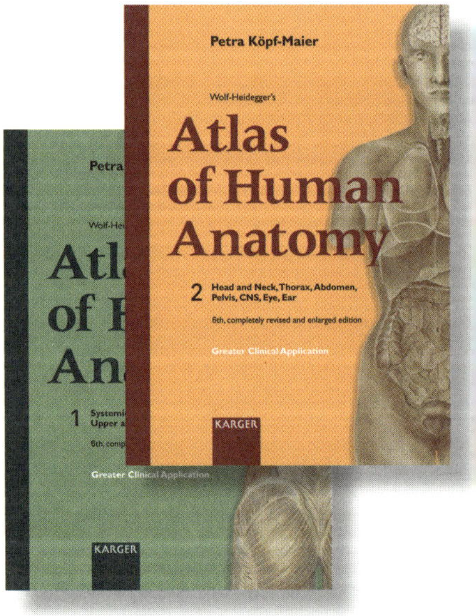

6th Edition 2005

However, the pathway to complete networking was long; many small steps had to be taken to prepare for digitalization. In the fall of 1990, the on-screen page proofs were already considered a great success. Maquettes – templates for typesetting on which individual photocopied elements, from text to tables and footnotes, were compiled – had outlived their usefulness. 'Assembly workers', who had previously pasted up these pages by hand, had to get used to working in front of computers. Understandably, the leap into the digital age was not easy for them, nor for many others. Familiar processes, some of which had been practiced for decades, had to be abandoned and new techniques learned. It was important for the company that digitalization would not cause bewilderment among the staff. Therefore, the management offered many training and further education courses in order to handle the transition together with the established employees.

Karger was well practiced in coping with technological upheavals; its affinity with IT solutions was developed early. In the early 1980s, every administrative workplace in the publishing house was already equipped with what were then called word processors, a cross between electronic typewriters and personal computers. Karger itself took over the training of employees and attracted the attention of IBM Switzerland, whose systems were used. IBM even commissioned Karger to take over training for IBM clients in the Basel area, including pharmaceutical companies.

At all levels, the publishing house gradually moved closer to the implementation of new electronic data processing solutions. It was the job of the computer coordination and network development team (*Informatik-Koordination und Netzentwicklung*, IKONE) to ensure that the computer science and information department developed practical solutions together with the staff. From spring 1991, IT coordinators in all departments set up a dialogue, and a little later the server went into operation with a 'company interfacing' concept.

The computer science and information department laid a solid technological foundation for the 'rationalization of operations'. The aim was to make work processes as simple, clear and economical as possible. Providing information to readers or users in digital form could not be foreseen at this stage. Like the rest of the publishing and media industry, Karger continued to rely on analog media. In 1991, the company's visual identity was again modernized. Uniform white and new typographic solutions then became the continuously updated standard for all journals. The awareness of the corporate and product design and the unique character of the publishing house was maintained – not least stimu-

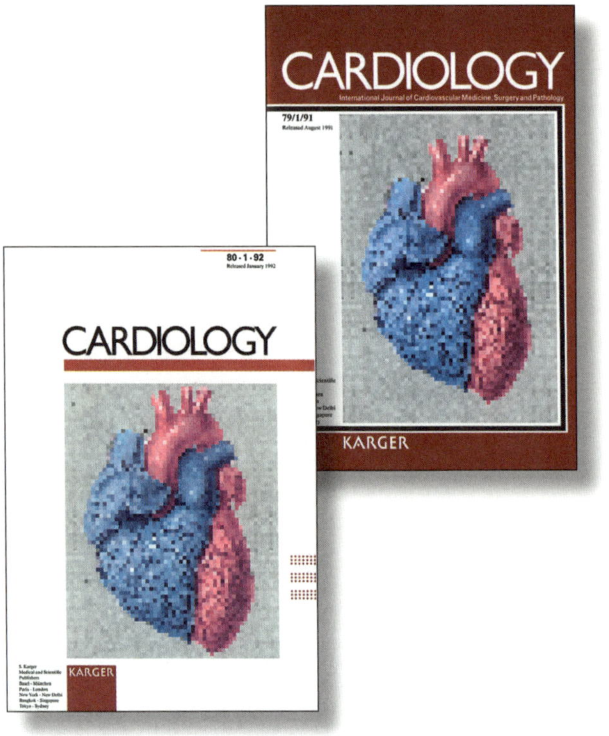

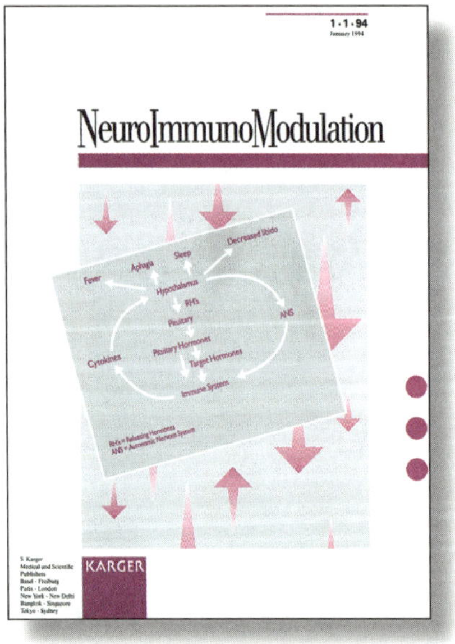

lated by the transformations in the scientific and publishing world. The more this was marked by concentration, the more clearly Karger recognized that it wanted to 'go its own way'.

The necessary changes to the company's electronic data processing system gave rise to an analysis coordinated by Steven Karger. As of 1992, all departments subjected their products and processes to a critical examination and defined subprojects. As a young publisher, Steven Karger put great emphasis on communicating all changes openly and transparently. Before the analysis, employees were informed that it would 'certainly keep us and you busy over the coming years'.

In 1992, the decision was made to reorganize the German branch. The branch in Germering near Munich had been an important component for sales on the German and Eastern European markets since 1969 and was the only branch to publish its own book and journal program. At the time, the move to Germany had been of great emotional significance for the publishing house after its displacement from Berlin, but after more than 20 years, 'Germering' was an integral part of the company. In the daily routine, however, its great distance from the headquarters in Basel often proved to be an obstacle. Therefore, in the spring of 1992, Karger Germany moved into offices in Freiburg, a university town just about 70 kilometers north of Basel. From there, the subsidiary continues to operate successfully today. The importance of the offshoot of the publishing house was revealed by the fact that Steven Karger also set up an office in Freiburg and from then on commuted between Freiburg and Basel.

Gabriella Karger, the younger daughter of Yvette and Thomas Karger, also joined the roughly ten-member team in Freiburg in October 1992. She remembers an occasionally turbulent period during which she nonetheless gradually began to enjoy the challenges the publishing industry presents. Like her

S. Karger GmbH, Wilhelmstrasse 20A, Freiburg, Germany

siblings, she had been familiar with everyday life in the publishing house from an early age. Now she was given the opportunity to shape it herself. Together with her brother Steven and the small team in Freiburg, she was able to try out many ideas. From here, Gabriella Karger's path led through many stages, some external to the publishing house, up to the top of the family business.

Strengthening the German branch proved a timely decision, because in December 1992, a very narrow majority of the Swiss electorate voted against joining the European Economic Area (EEA). Therefore, the easier access to the European markets that Karger – as many export-oriented Swiss companies – desired was initially denied. Karger particularly had to look for different channels for sales and distribution. In addition to the libraries and specialist retailers in Germany, which were among the most important customers, Eastern European countries required special attention. Contact with many customers there was lost after the dissolution of the communist states. Karger's strategy was to be present in these markets with a separate branch within the EEA. Freiburg would be the base for maintaining and developing contacts in Germany and to regain Eastern European markets.

Thinking and planning in a more market-oriented way was, of course, also essential for the executive staff in Basel. The publishing house had to abandon its inside focus and view itself from an outside perspective. The organization was still completely focused on producing books and journals; the editorial and production departments worked according to the requirements of these two product groups, regardless of the subject matter. But in the daily routine, direct contact with researchers in the medical fields was the most important aspect. After all, Karger published information for fields such as nephrology, endocrinology and genetics. In these and other increasingly specialized disciplines of medicine, it had to find editors, build and main-

tain a network and develop market-oriented publications. Shouldn't the publishing house structure itself internally along the same lines? In the spring of 1993, after intensive discussions, the decision was made to begin drawing up a new divisional organization according to medical fields.

Thomas Nold, the head of the editorial staff, advocated defining fields of expertise in which product and marketing managers supervised the development from aquisition to the final publication. In this way, they would become known and valued partners for scientists and thus represent the publishing house. Many at Karger felt this move away from the structural breakdown by type of publication – journals and books – to focusing on specific disciplines was a major turning point. Karger thus took another big step towards becoming a modern team-based media company.

Thomas Karger after having been awarded the honorary doctorate in the Martinskirche Basel

The significance of the publishing house and in particular of the publisher Thomas Karger for scientific research was confirmed in November 1993 when Thomas Karger received the honorary doctorate from the Medical Faculty of the University of Basel. The laudatory speech said that by founding and publishing numerous renowned professional journals, Karger raised the reputation of the university as well as of Basel as a science center. The organization of scientific symposia and the promotion of young scientists also received high recognition in the awarding of the honorary doctorate. For Thomas Karger, it was the second award of its kind after the honorary doctorate awarded to him by the Hamburg Faculty of Medicine in 1972. Thomas Karger felt it was a special honor to receive it in Basel, the home of the publishing house. However, he stressed that, in addition to his wife and family, the staff and the researchers published by his house had contributed substantially to the publisher's performance.

The program and the content of Karger remained essentially unchanged, but the range grew immensely in the 1990s. Not least to cope with this increase in publications, in particular the new journals, Karger required more efficient operations, and data processing offered the needed support at all levels. The publishing house proved versatile while retaining staff continuity – an important signal both internally and to researchers in medicine that were associated with Karger as authors, editors and readers.

There was still uncertainty about the future of electronic publishing. Thomas and

Steven Karger knew that they wanted and needed to exploit the opportunities offered by the new media. For some time, they attentively examined the best ways to successfully access this market and held discussions with potential partners. There were hundreds of possibilities that not only involved uncertainties but also required major investments and cooperation with external service providers.

When, in 1995, Thomas Karger made the fundamental decision to create a website and commence electronic publishing, he was happy about his initial reluctance. Unlike other publishers, Karger had 'not stumbled into it' and had not spent large sums of money on poor investments. He took the pathway into the World Wide Web with the help of Vitek Tracz, who had founded the media and service company BioMedNet in London. Thomas Karger had met and come to trust the Polish-born entrepreneur; he was the first person with whom he could imagine collaborating. Tracz had the necessary expertise and had 'now surely tested' the potential of the Internet 'as he has shown us', Thomas Karger said to convince his management team. 'So we won't be running after some theory.'

In Basel, Karger compiled the data for the website and delivered them to London, where they would be posted online. 'At the moment, this is all still so new that we are diligently working to gain experience', reported Steven Karger in a lecture, 'the technological tools exist and are being applied step by step.' Yet, at the publishing house, the mood continued to sway between euphoria and skepticism with regard to electronic media. Above all, Thomas and Steven Karger, like many entrepreneurs from the publishing and media industry, wondered how their services would be marketed in the future. Should journal subscribers receive access to electronic publications as an alternative or in addition to the printed product? Were the users' and Karger's own networks sufficient for exhausting all possibilities? Basically, there was no avoiding electronic publishing; however, it could still take some time until it made any money.

In the summer of 1996, the publisher's first website went online at www.karger.ch. The time of reflection and deliberation ended here; the first website was a milestone on the road to Karger's digital future. The website started by guiding users to an alphabetical list of publications in the relevant fields. New titles were presented in a 'News' section, and the tables of contents of journals could be found on the web several weeks before publi-

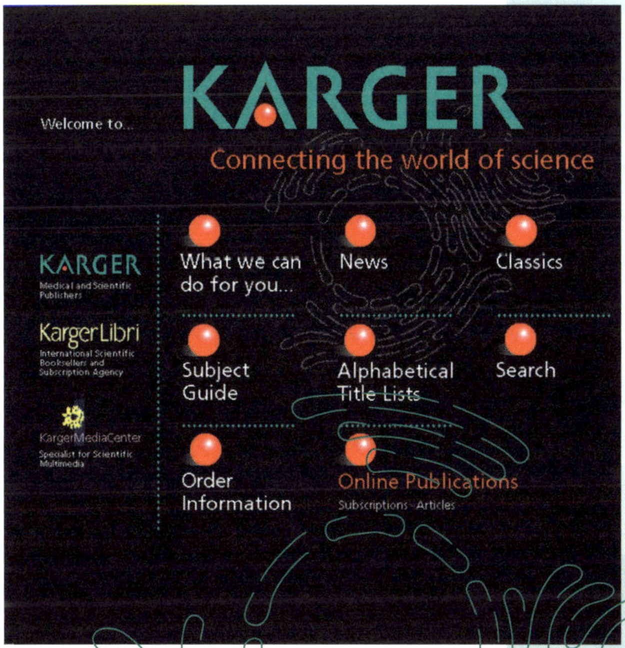

cation. Even full-text issues of the *European Journal of Human Genetics* were accessible on the site via a link to BioMedNet. In addition, it was possible to contact the publisher via the website, whether about publications, advertisements, or customer services. In the next step, users would be able to order journals – initially only print versions – directly via the website.

Soon enough, digitalization of the publishing routine sped up considerably. New typesetting software and improved exchange of data between departments also generated many innovations. At the same time, Karger was looking for its own online solution. Not organizing this central area of the publishing business on its own did not correspond with the publishing philosophy of Karger.

In 1997, Karger put its entire program online – at the same time as its major competitors. Impressively, it proved that a smaller, privately operated company could move at the same high technical level as the industry giants. In Basel, the company carried out the carefully prepared consolidation of the data. At the new address www.karger.com, Karger began, for the first time, offering online publications on the Internet. The site was relaunched on this occasion for the first time. Ease of use was a primary concern, because the site was not only seen as a medium for communication, but primarily as a basis for e-commerce. While users previously could only purchase single articles online, journal subscriptions were also available from 1998 onwards.

The environment for scientific publishing had changed slowly but palpably in the 1990s not only through the electronic media. 'As in other industries, companies are being bought up and taken over, generating impersonal mammoth companies', Thomas Karger described this development to the staff. In particular, two major media and publishing groups gained increasingly dominant positions in the field of medicine – with consequences for the publisher's work, as he added. 'The personal supervision of disciplines in medicine and personal contact with authors over years is being rationalized'. At the same time, this tough competition created interesting niches for publishers like Karger. 'If the aspect of individuality, of the interpersonal relations, can still be lived by a publishing house, this is one of the rare opportunities for it to differentiate itself favorably from the uniformity and sterility of the competition'. Together with his family and staff, the publisher wanted to continue to raise this profile. Long experience, great expertise, and reliable and personal service were qualities that led Karger to believe that it would be able to continue to perform well.

In spite of all its worldwide activities, Karger remained firmly anchored in Basel. In addition to the publishing headquarters at Allschwilerstrasse, in the old town – very close to the university – Karger operated an international subscription agency and a bookshop, both under the name Karger Libri. The bookshop, called 'Haus des Wissens' (House of Knowledge), was renovated by Steven Karger in 1997. A café in the front of the store where students could grab a bite to eat was established and run for some years opposite the main building of the university. In addition to the changed appearance the digitalization also asked for a change in focus. The bookstore offered eBooks through its website, and a web shop was added. However,

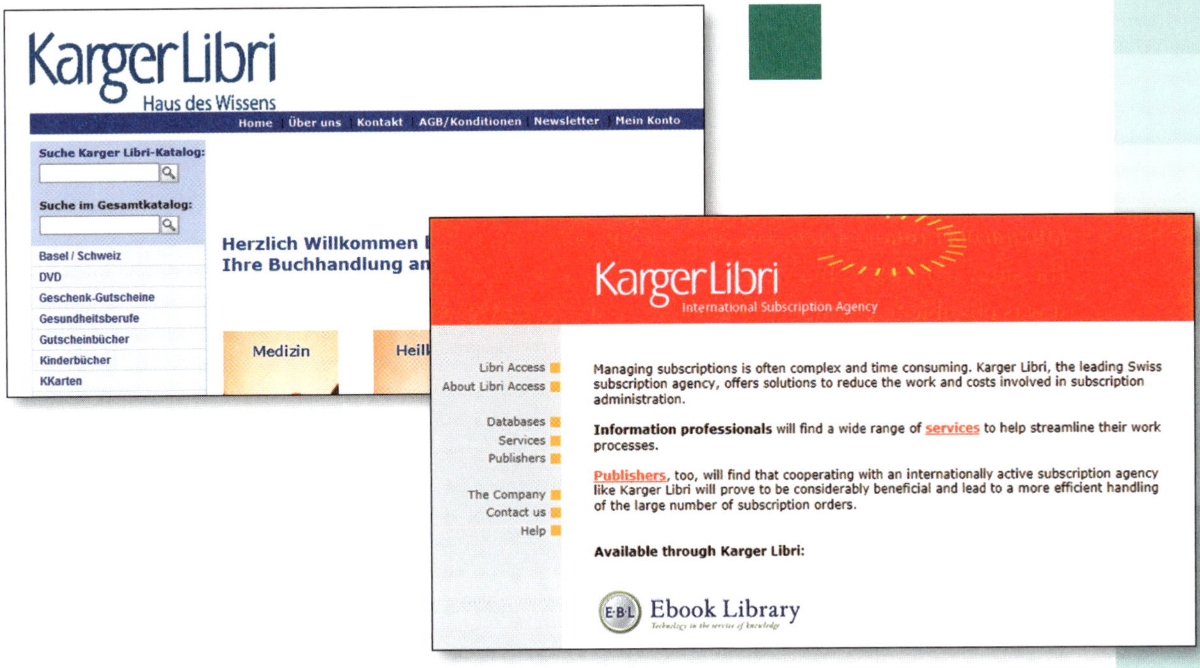

the bookshop was increasingly challenged by the competition of the big online retailers, and the reading habits of students – one of the most important customer groups – also changed.

Digitalization also meant major changes for the subscription agency. Nonetheless, the customers, in particular the institutional libraries, continued to make use of a distributor. It was too complicated to remain in contact with each individual publisher, for example to record and check IP ranges. The Libri subscription agency had already operated branches in Japan and Thailand since the 1980s. When the Japanese government decided to raise taxes for foreign companies in the domestic market in 2012, the office in Japan left the subscription business. It focused on databases and other products and only employed one staff member on site. The Bangkok office, in contrast, flourished and could be expanded in 2014. When one of the major competitors in the journal subscription business announced bankruptcy in September 2014, Karger Libri's subscription agency was able to seize the chance and win new customers in various markets. How this business area will evolve is, however, an open question, because this insolvency sparked debate about the need for and the benefits of distributors.

The pressure of competition in the international scientific and publishing world has grown since the early 1990s, as decades-long development reached its limitations. For many decades, authors and editors, readers, libraries and publishing houses had become used to more and more books and journals being produced, sold, purchased and made available to users. This came to an end. Everywhere in Europe and the United States – the most important markets for Karger publications – libraries and institutions cut their budgets. Although his publishing house was yet spared from economic problems that beset other industries worldwide, as Thomas Karger stressed, it was also affected by declining subscriber numbers and book sales.

The budget cuts were the starting point of a ruinous development: academic publishers

'Compiling, viewing, selecting, condensing and disseminating information' remains the genuine mission of the publishing house.

tried to compensate for lost sales by increasing their prices; especially the large media corporations with their publications that were indispensable for the institutions began aggressive pricing policies. Rapidly rising prices, however, led to further cancellations of journal subscriptions – the notorious 'serials cancellation crisis' began spiraling ever faster. Often, higher production costs caused price increases also for digital publications, but some publishing houses simply took advantage of their dominant positions. And at the same time, the relationship with libraries transformed as they became the most important customers. Again and again, costly subscriptions exceeded the tight budgets, resulting in tough negotiations during which all publishers had to make concessions. Unmistakably, the seller's market had finally become a buyer's market.

During the years of digitalization, the lack of understanding among many buyers and subscribers for pricing was combined with fundamental doubts about the service of publishing. Did science really still need publishers as brokers of knowledge? Why not self-publish results? During the serials cancellation crisis, the idea of open access to scientific data on the Internet took root. However, like many other publishers, Thomas and Steven Karger were convinced that even – and in particular – electronic publications needed the editorial and technological support of experienced publishing houses. Their role in the scientific system was increasingly questioned. Nevertheless, Karger took on this challenge.

It was now first and foremost Steven Karger's job to guide the publishing house established by his great-grandfather into the 21st century. He felt strongly committed to the tradition established by the previous generations: this meant that Karger remain a specialist for publications of the highest scientific quality. 'Compiling, viewing, selecting, condensing and disseminating information' remained the genuine mission of the publishing house – even if it frequently was carried out with the help of electronics and resulted in digital media. It was now time to profitably link the traditional strengths of the publishing house with the expansion of electronic publishing. In the second half of 1999, Steven Karger took over from his father as CEO and embarked upon this mission together with his management team.

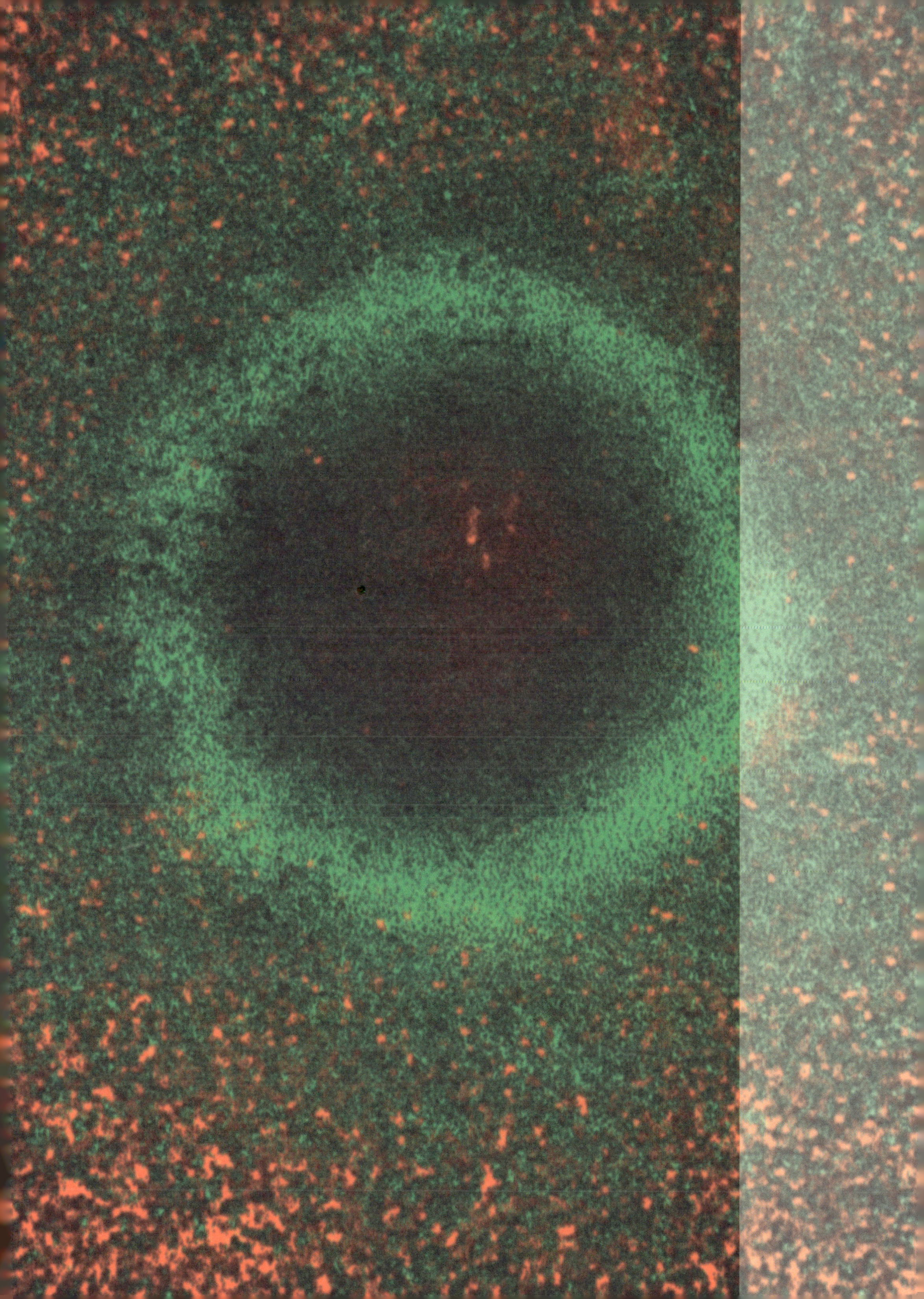

8

The Fourth Generation

KARGER

Steven Karger

'I am taking over a good, intact, sound and future-oriented business', Steven Karger wrote to the staff of the publishing house at the beginning of 1999, 'and I will do everything in my power to make sure it stays that way.' He added that the changes in the world of science and the publishing industry and the concentrations and mergers had their impact on the business but did not fundamentally challenge the S. Karger company. With its lean structure, the publishing house was large enough to stay competitive in the market, and yet small enough to operate flexibly. 'We can build on our strengths and continue our niche policy, which has worked out very well for us so far and is ultimately the key to our success', Steven Karger stated. Versatile, efficient and geared to the needs of researchers – these were the assets of the publishing house. At Karger, decisions were made quickly and straightforwardly, whereas in large corporations, they were dependent on many factors and people. Rather than having the company's long-term developments in mind, they were aimed at quick returns on investment.

Karger repeatedly received proposals for cooperation or takeover from large publishing houses, which did not go unnoticed by the staff when meetings and negotiations took place. However, the family never questioned its financial independence: 'We won't sell.' For this reason, Steven Karger talked openly about offers and even published the verbatim letter of refusal to an interested US publisher in the spring of 1999 in the staff newsletter, stating that the company definitely wished to maintain its independence, not least due to a feeling of responsibility for its employees. 'If there was a change of ownership, we could no longer meet this responsibility and would only have limited influence on preserving existing jobs', Steven Karger said, sending out a clear signal to the industry and the company's employees.

Being a representative of the fourth generation, Steven Karger was determined to follow the policy of his father, grandfather and great-grandfather in his commitment to the independence of the family business. He also adhered to the fundamental idea of the publishing house as a broker of knowledge for medical research. Steven Karger saw his challenge in advancing electronic publishing and making it a central pillar of the business. In 1999 and 2000, the company was already deeply immersed in the process, but there were still major steps ahead of everyone involved. This also required a 'change of mindset'. 'Perceiving the new medium as an opportunity must be encouraged', Steven Karger pointed out.

The year 2000 was marked by projects of process automation and the introduction of new software, which affected all departments from the editorial to the production and delivery teams. Staff members worked tirelessly in project teams to define service catalogs and configure and evaluate programs. The implementation of the new technologies was linked with far-reaching strategic decisions. Making contents available in electronic media was only part of the task. Marketing them was much more difficult in this period of fundamental change, given that many questions were still unanswered: What did customers expect, for example, with regard to traceability and supplementary information? What did competitors offer? What standards were evolving in the international scientific community? And how long was the life of a given digital solution?

To gain confidence in this complex business segment, Karger put much effort into achieving major milestones. This included a product database suited for all purposes from deliveries via the online catalog to the supply of data to external partners. It took shape in 2001, so that data – from the names of the journal editors to the ISBN and ISSN numbers – have ever since been administered directly and by automated means and no longer need to be tediously fed into each single system. After a period of extensive editing and expansion, vast amounts of data were available, including, for example, bibliographic data dating back to 1890 as well as the titles of all articles and all authors' names from the previous years. According to Mike Dürst, head of the Publishing and Information Technology Division, the dynamic link established between the central data source and the website was a turning point. While individual interventions were possible, they were to remain the exception. 'Putting the reach into research' was the motto behind the transmission of data using as much automation and as little manual intervention as possible to dozens of specialized scientific research, discovery, and Abstracting & Indexing platforms on a global scale.

'Putting the reach into research' was the motto behind the transmission of data using as much automation and as little manual intervention as possible.

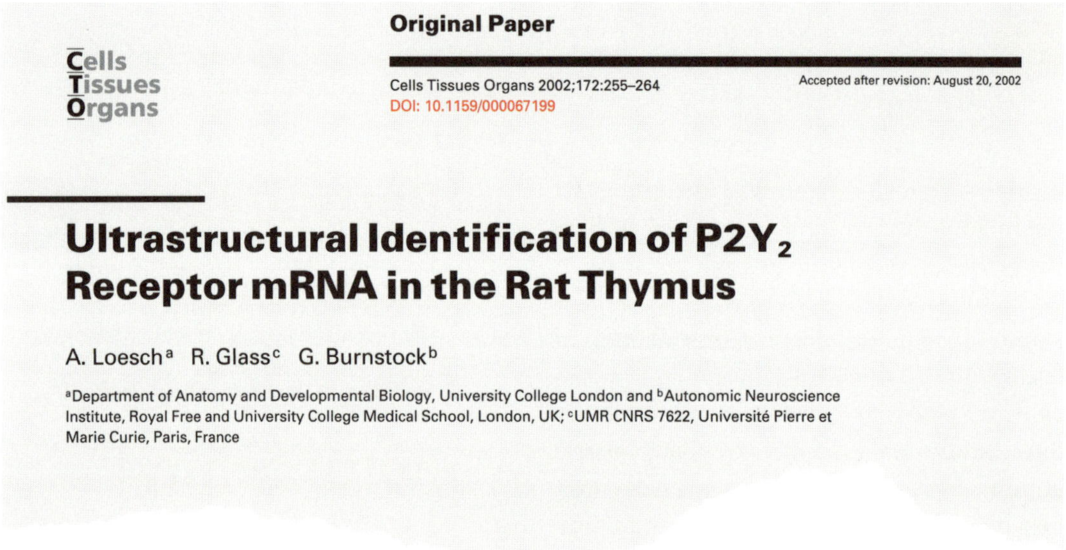
Original Paper

Cells Tissues Organs

Cells Tissues Organs 2002;172:255–264
DOI: 10.1159/000067199

Accepted after revision: August 20, 2002

Ultrastructural Identification of $P2Y_2$ Receptor mRNA in the Rat Thymus

A. Loesch[a] R. Glass[c] G. Burnstock[b]

[a]Department of Anatomy and Developmental Biology, University College London and [b]Autonomic Neuroscience Institute, Royal Free and University College Medical School, London, UK; [c]UMR CNRS 7622, Université Pierre et Marie Curie, Paris, France

The digital object identifier (DOI): a character string used to uniquely and persistently identify an electronic document

At that time, there was no way of anticipating the significance of this milestone in the publishing company's development, because online services were still regarded primarily as an indispensable supplement to the print editions. The use of online services, however, quickly picked up speed, especially since Karger kept adding new features to its website. The article references were linked to meta-databases such as PubMed, articles were made identifiable based on the new DOI (digital object identifier) standard and additional functions made searches on the website easier. Libraries, the main customers, were interested in advanced statistical solutions about usage figures to content.

Within a few years, the new forms of publishing changed the users' behaviors. Some customers could already do without printed versions and, starting in 2003, could opt for 'online only' subscriptions to journals. The access figures recorded for individual articles skyrocketed to more than one million in the same year. Journals such as *Cytogenetics and Genome Research* took leading positions: an article from this publication was accessed on average once every ten minutes, which seemed highly remarkable at the time. The opportunity to select specific information heralded a new era. Journals and books continued to be collection boxes for articles and chapters, and often only single sections of them were used. The advances in the electronic use made more fine-tuning possible. At the beginning of the digital age, just as in the days of print, it was still necessary to leaf through an entire issue of a journal – even if only interested in one specific article. It did not take long until articles could be accessed directly without dealing with the other content. However, this entailed a progressive loss of 'serendipity' – the discovery of information not explicitly searched. This phenomenon, which arose with the improved search options of the Internet, is being hotly debated in many scientific and specialist communities today.

By the turn of the millennium, researchers around the world used the Internet as a matter of course. They retrieved information on the websites of the publishers, and, starting in 2002, Karger have also offered authors the possibility of submitting manuscripts online. Authors need to ensure that certain standards are already met at submission, such as the proper configuration of references. Compliance with such standards facilitates publishing – both for publishers and authors. In addition, online procedures have also been applied in the peer review process. Article reviews – essentially a communication process between journal editorial departments worldwide and authors and reviewers on all continents as well as with the publishing house in Basel – cannot only be completed faster, but also in a more transparent way and, in particular, with better scientific quality.

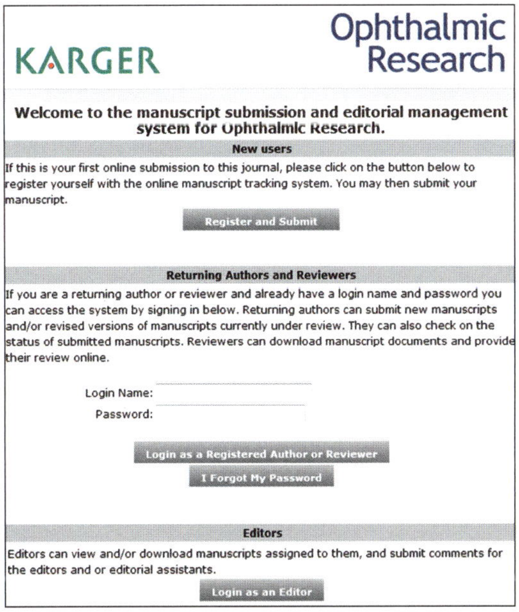

Online manuscript submission and peer review – a standard since 2002

The online publishing segment perceptibly gained momentum, although in 2003 it still accounted for only 10% of the publisher's total turnover, as Steven Karger warned. He said it would still have to grow substantially in the upcoming years, not least for Karger to stay competitive. The editorial, information technology and sales teams worked hard to achieve further improvements. While Karger had been able to rely on worldwide library subscriptions to its journals and book series for decades, it was now faced with new challenges. The publishing house made substantial efforts to implement efficient invoicing software for the various online products. In addition, Karger stepped up cooperation with distributors and journal agencies that not only handled billing, but also marketing and communication tasks, since by now smaller publishing houses like Karger had to outsource services to strengthen their own sales network. With these partners, Karger succeeded in highlighting unique specialities of its program and bringing them to the global marketplace.

As prices increased, large libraries in particular expected more services from the publishing houses in addition to pure and simple information. This mainly included supplementary service offers: libraries updated their catalogs with bibliographic data in MARC (Machine-Readable Cataloging) format, and the SUSHI (Standardized Usage Statistics Harvesting Initiative) standard was used for fully automated collection and statistical processing of electronic media usage data. After all, the libraries themselves were under pressure. As budgets were tight, they had to gain insight into content usage to assist them in selecting publications.

It was essential to step up sales activities, given that negotiations became more difficult as the serials crisis escalated. For this reason, Karger, like all scientific publishers, established close links between sales and its own IT team to meet expectations. Indexing information and making it permanently traceable remains a major task.

Since 2000, technological advances and changes in the international publishing landscape have begun setting the entire STM industry in motion. One of the major topics was open access – i.e. unrestricted access to scientific information on the Internet. When research-funding agencies started providing research grants only if the results were published with free access, publishers eventually began to consider this model in depth. Whether open access would be 'friend or foe' was difficult to predict, Steven Karger told his employees in 2005. While policymakers would usually clearly favor it, authors were still cautious when it came to paying a publication fee. Readers or users who gained access to the information from the websites of their institutes or university libraries, by contrast, showed hardly any interest in whether access was free or based on a paid subscription.

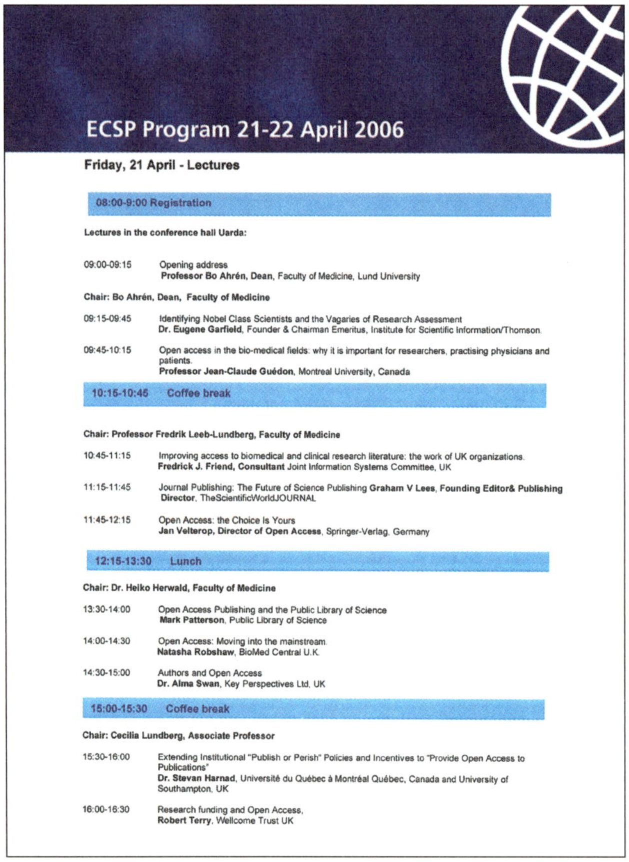
ECSP Program 21-22 April 2006

Friday, 21 April - Lectures

08:00-9:00 Registration

Lectures in the conference hall Uarda:

09:00-09:15 Opening address
Professor Bo Ahrén, Dean, Faculty of Medicine, Lund University

Chair: Bo Ahrén, Dean, Faculty of Medicine

09:15-09:45 Identifying Nobel Class Scientists and the Vagaries of Research Assessment
Dr. Eugene Garfield, Founder & Chairman Emeritus, Institute for Scientific Information/Thomson.

09:45-10:15 Open access in the bio-medical fields: why it is important for researchers, practising physicians and patients.
Professor Jean-Claude Guédon, Montreal University, Canada

10:15-10:45 Coffee break

Chair: Professor Fredrik Leeb-Lundberg, Faculty of Medicine

10:45-11:15 Improving access to biomedical and clinical research literature: the work of UK organizations.
Fredrick J. Friend, Consultant Joint Information Systems Committee, UK

11:15-11:45 Journal Publishing: The Future of Science Publishing **Graham V Lees, Founding Editor& Publishing Director**, TheScientificWorldJOURNAL

11:45-12:15 Open Access: the Choice Is Yours
Jan Velterop, Director of Open Access, Springer-Verlag, Germany

12:15-13:30 Lunch

Chair: Dr. Heiko Herwald, Faculty of Medicine

13:30-14:00 Open Access Publishing and the Public Library of Science
Mark Patterson, Public Library of Science

14:00-14:30 Open Access: Moving into the mainstream.
Natasha Robshaw, BioMed Central U.K.

14:30-15:00 Authors and Open Access
Dr. Alma Swan, Key Perspectives Ltd, UK

15:00-15:30 Coffee break

Chair: Cecilia Lundberg, Associate Professor

15:30-16:00 Extending Institutional "Publish or Perish" Policies and Incentives to "Provide Open Access to Publications"
Dr. Stevan Harnad, Université du Québec à Montréal Québec, Canada and University of Southampton, UK

16:00-16:30 Research funding and Open Access,
Robert Terry, Wellcome Trust UK

Program of the 1st Conference on Scientific Publishing in Biomedicine and Medicine, April 21–22, 2006 in Lund, Sweden

Karger first primarily observed the development of the model and in 2005, decided to provide free access only to individual articles or collections to begin with, arguing that it was not to be expected that subscriptions would be displaced, as open access might rather be one publishing option alongside others in the future. During a conference on scientific publishing in 2006, Steven Karger gained the impression that one also had to leave it to the authors as to whether they could and wanted to put money into their publications. The one thing that was clear to him was that the services of the publisher were indispensable, especially in electronic publishing. The question of who would pay the costs would perhaps in the future be of secondary importance or decided from case to case.

The louder the calls for free publications and the tighter library budgets became, the more important were the selection criteria.

ISI Web of Knowledge℠

Journal Citation Reports®

WELCOME | HELP | 2013 JCR Scie[illegible]

Journal Summary List

Journals from: subject categories PSYCHIATRY | VIEW CATEGORY SUMMARY LIST

Sorted by: Impact Factor | SORT AGAIN

Journals 1 - 20 (of 136) | [1 | 2 | 3 | 4 | 5 | 6 | 7]

MARK ALL | UPDATE MARKED LIST

Ranking is based on your journal and sort selections.

Mark	Rank	Abbreviated Journal Title (linked to journal information)	ISSN	JCR Data: Total Cites	JCR Data: Impact Factor	JCR Data: 5-Year Impact Factor	JCR Data: Immediacy Index	JCR Data: Articles	JCR Data: Cited Half-life	Eigenfactor® Metrics: Eigenfactor® Score	Eigenfactor® Metrics: Article Influence® Score
☐	1	MOL PSYCHIATR	1359-4184	13902	15.147	14.196	3.500	132	5.4	0.03924	4.672
☐	2	ARCH GEN PSYCHIAT	0003-990X	38622	13.747	14.396		0	>10.0	0.05448	5.215
☐	3	AM J PSYCHIAT	0002-953X	43193	13.559	15.062	3.361	119	>10.0	0.04917	5.204
☐	4	WORLD PSYCHIATRY	1723-8617	1459	12.846	8.093	1.842	19	4.3	0.00377	2.115
☐	5	BIOL PSYCHIAT	0006-3223	40687	9.472	10.347	2.883	239	7.4	0.08145	3.332
☐	6	PSYCHOTHER PSYCHOSOM	0033-3190	3073	9.370	7.024	1.633	30	7.5	0.00555	1.961
☐	7	SCHIZOPHRENIA BULL	0586-7614	13188	8.607	9.222	2.049	144	6.4	0.02941	2.775
☐	8	NEUROPSYCHOPHARMACOL	0893-133X	[illegible]	[illegible]	[illegible]	1.765	255	[illegible]	[illegible]	2.820

One of these criteria is the impact factor (IF) – an index that reflects the number of citations of an article from an academic journal in other publications. However, lately, publishers have observed with increasing concern that the IF is no longer used only as a bibliometric value. It has also become a criterion for the recruitment of researchers at universities and the allocation of grants in research funding. 'What was intended to be a simple quality index has become a phenomenon that dominates everything', said Thomas Karger in 2006. Together with Eugene Garfield, the founder of the American Institute for Scientific Information (ISI), which evaluated the literature databases, he himself had argued early on for a quantitative assessment of the quality of scientific papers and journals. When this idea came up in the 1960s, the IF seemed to be a valuable instrument, especially for publishers like Karger focused on high quality. The IF often provided evidence of the great response to and impact of the articles in Karger's journals. The idea was good, Thomas Karger still believed decades later. 'None of us could picture the consequences this would eventually entail.'

The IF, as a supposedly representative value, is associated with many methodological problems from the period of assessment to researchers citing themselves and many more. Even if the IF continues to be a viable assessment tool in major fields of medicine, it has proven to be not very significant in many more specialized areas. Qualitative aspects carry too little weight in the analysis and, first and foremost, the IF can be manipulated. However, as it is one of only few measurable indicators, it continues to have major influence and both researchers and publishers must live with. Although the IF is controversial, it remains a fixed part of the medical science publishing world; it has become a measure that is difficult to calculate, also for Karger. It is nevertheless indispensable to consult the Citation Index to estimate the

Karger offices and representations around the globe

value and development of one's own publications. Notwithstanding, it has been, and continues to be, Karger's policy to prove its quality beyond the IF. Personal contacts, publishing excellence and deep-rooted quality awareness have been and continue to be essential factors that have preserved and increased Karger's reputation to this day.

Concerning the distribution of its own products, Karger continued to still see potential for development based on personal relations after 2000. The contacts with large libraries in the key markets of the United States, Europe and Japan, some of which had existed for decades, were a major benefit in this context. Its branches in the United States (S. Karger Publishers, Inc., Farmington, Connecticut, since 1979) and in Japan (Karger Japan, Inc., Tokyo, since 1981) continued to be highly significant and were primarily in charge of sales and marketing.

The Karger family also established contacts in other Asian regions early on. Moritz Thommen, head of the Sales Division, used these contacts to drive the international business and opened additional branches. In 2005, Karger founded its Karger Regional Office for Asia in Shanghai, China. It moved to Kuala Lumpur, Malaysia, in 2010, and now runs the sales activities in China, Taiwan, Malaysia and Southeast Asia (excluding Japan and Korea). S. Karger Thailand was established in 2010 in Bangkok, Thailand, where Karger already operated a branch of the Karger Libri subscription agency, to open up the scientific market. In India, the Karger Sales Department had been working for many years very closely in succession with two different local partners. The cooperation continued as of 2012 under the umbrella of Karger India Private Ltd.

To successfully gain a foothold in the emerging Latin American market, a pre-existing collaboration with a Brazilian agency was strengthened by employing an associate exclusively representing Karger publications. Karger continues to view agencies as good partners to achieve presence, for example in Australia, the Middle East and many other regions. Karger has appointed a local representative in Africa, an approach that has been successfully applied in Pakistan for many years.

In addition, in 2012, Karger founded the Editorial Division Asia (EDA) as a central hub for Asia. Led by Peter Roth, head of Editorial Deptartment and Services, staff members in Japan, China and India have since then been assisting in the procurement and management of English-language publication projects in these parts of the world. The idea of this approach is to build a bridge between Asia and Europe and the United States.

It is evident that the economic power of, for instance, Korea has been increasing. Research activities are being reinforced in many Asian countries, stimulating the growth of the market for scientific publications. The scientific community in Asia makes significant contributions to international medicine and is on the lookout for media to publish them.

Personal relations with the publishing house have been and continue to be an essential part of international activities. This approach has been pursued with close partners in the major scientific nations in Europe, the United States and Japan. However, new customers in Asia, South America or emerging science nations such as Turkey also appreciated a personal presentation of the company's publications by a member of the Karger family or senior management. As it had been done for decades, scientific conferences have been and continue to be attended and institutes or university libraries have been and are being visited to seal contracts with groups or to market unique, customized solutions.

Ever since libraries have merged to form consortia at the local or regional level or even negotiated with publishers on the basis of national licenses, they have been able to provide a wider range of publications including the STM sector. This has saved both libraries and publishers the trouble of many individual negotiations. Consortia further offered publishers like Karger an opportunity to find long-term customers for their entire publishing program. The sales teams realized major consortial deals, for instance in the important Asian markets, in California, Brazil or many European countries. In Germany, Karger negotiated a national license, which granted all German universities access to the Karger journals. However, this model entailed the risk of having to offer discounts too big to make it worthwhile. Therefore, prudent calculations and planning were the keys to this approach. While many libraries continued to

Homepage of Karger China (www.karger.cn)

subscribe to the big-name titles, the situation gradually became more difficult for more specialized publications. To protect them, Karger formed collections and packages of publications having certain elements in common. While some libraries wished to acquire the entire package of Karger publications at an attractive price, others were interested in a specific selection, for example in a given medical specialization.

To complete its range of publications available online, Karger also took on the task of retro-digitizing its journal contents in 2006. This was followed by the recent completion of digitalization of all book publications. Altogether more than 2 million pages of content are now available. The long-standing publishing house possessed a huge collection of older publications that were of major interest for some users, and it had the technological means to recover these treasures with reasonable effort. It soon became apparent that there was a great demand for subsequently digitized information in some mar-

Acta Cytologica
Acta Haematologica
Nephrology
Nutrition & Metabolism
Audiology Neurotology
Blood Purification
Brain, Behavior and Evolution
Breast Care
Cardiology
CardioRenal Medicine
Caries Research
Cellular Physiology and Biochemistry
Cells Tissues Organs
Cerebrovascular Diseases
Chemotherapy
Current Urology
Cytogenetic and Genome Research
Dementia and Geriatric Cognitive Disorders
Dermatology
Dermato pathology
Developmental Neuroscience
Digestion
Digestive Diseases
Digestive Surgery
European Addiction Research
European Neurology
European Surgical Research
European Thyroid Journal
Fetal Diagnosis and Therapy
Folia Phoniatrica et Logopaedica
Folia Primatologica
Food Science and Technology Research
Forschende Komplementärmedizin
Gastro Intestinal Tumors
Gerontology
Gynecologic and Obstetric Investigation
Hormone Research in Pædiatrics
Human Development
Human Heredity
Allergy and Immunology
Integrative Medicine International
Interventional Neurology
Intervirology
Journal of Innate Immunity
Journal of Molecular Microbiology and Biotechnology
Journal of Nutrigenetics and Nutrigenomics
Journal of Vascular Research
Kidney Blood Pressure Research
Kompass
Liver Cancer
Medical Epigenetics
Medical Principles and Practice
Molecular Neuropsychiatry
Molecular Syndromology
Neonatology
nephron
Neurodegenerative Diseases
Neuroendocrinology
Neuroepidemiology
NeuroImmunoModulation
Neuropsychobiology
Neuro-Signals
Obesity Facts
Ocular Oncology and Pathology
Oncology
Oncology Research and Treatment
Ophthalmic Research
Ophthalmologica
ORL
Pathobiology
Pediatric Neurosurgery
Pharmacology
Phonetica
Psychopathology
Psychotherapy and Psychosomatics
Public Health Genomics
Pulse
Respiration
Ganzheitsmedizin
Sexual Development
Skin Appendage Disorders
Skin Pharmacology and Physiology
Stereotactic and Functional Neurosurgery
Transfusion Medicine and Hemotherapy
Urologia Internationalis
Verhaltenstherapie
Viszeralmedizin

kets and on certain topics. Thereby, the publisher also met the demand for abundant and diverse scientific information.

While Karger was happy in its role as a rather small player, it also endeavored to place its business on a broad basis and to become noticed. It was clear that only those who delivered compelling quality and quantity would survive in the international publishing landscape and prevail against competitors. At this point, Karger had long been not only marketing its relatively constant portfolio of somewhat more than 80 journals and 40 book series, but also a wealth of information for a wide range of demands.

The program managers at the publishing house continued to face the challenging task of recruiting the right editors and acquire the right publications. There was no lack of material when it came to procurement, as more and more nations contributed to research worldwide and augmented both the scientific community and the demand for publishing. The main challenge for a scientific publishing house now was to identify the right topics and experts among the huge variety at hand. For example, while the field of classic anatomy lost in interest, the relevance of molecular biology and genetics and also fetal medicine and gerontology grew. The publishing program had to take account of the major medical challenges, such as dementia or cancer, and personalized medicine. Although making out trends, winning experts in the related fields and organizing a publication with them had been one of Karger's strengths for a long time, those in charge for the publication program experienced an ever-increasing level of specialization and observed shorter life cycles of the publications. More and more, researchers depended on a quick publication of their results. They felt the increasing pressure from their sponsors to continuously publish their scientific progress.

Today's headquarters of Karger Publishers, Allschwilerstrasse 10, Basel

Developments in science confronted Karger with new questions. Should it continue to pursue the goal of publishing information for most fields of medicine? Did the publishing program not already reflect certain specializations? While such topics were debated, one unique aspect of the publishing house became increasingly evident: the highly personal relations it had entertained with editors and authors for generations. As the research landscape became more complex, researchers increasingly appreciated reliable partners in a publishing house. Karger stood out favorably with its experienced editors who were familiar with international medicine and flexible enough to meet special publication needs. This company culture was now upheld especially by Steven Karger, along with Thomas Karger and long-standing members of his staff. Together they maintained contacts with researchers around the world and promoted the development of electronic publishing from their Basel headquarters. Steven Karger had become an important person for many employees to approach with ideas and their implementation.

When Steven Karger was diagnosed with a brain tumor in November 2006, it came as an unfathomable shock for his family. The employees of the publishing house were also shaken when they learned of the diagnosis. The disease had a profound impact on the company, given its structure as a family business. At the end of the year, Steven Karger addressed the staff members personally, thanking them for the many getwell wishes. 'I will keep on fighting, continue to be in the company regularly and carry out my duties', he promised.

The activities in electronic publishing, initiated and advanced by Steven Karger, continued to require major input. The employees had their hands full with new technological solutions, new products and additional services. In addition, the company premises were expanded. Construction work on a new building began in 2007 on the property at Friedrichstrasse, which has been in use since 2009. The development of the publishing house offered every reason for optimism, but was overshadowed by the illness of Steven Karger. During 2007, it became increasingly clear that he could not return to the company. Steven Karger himself and his father Thomas Karger were considering the future and turned to Steven's younger sister, Gabriella Karger. She had held various editorial positions in the company until 2001, but felt she wanted to stand her ground again outside the family business. Therefore, she switched over to another medium and took on a position with the cultural channel of the Swiss Broadcasting Corporation. Since her successful cooperation in Freiburg, she had known that publishing work is very exciting and varied. Gabriella Karger saw the publishing house as an interesting challenge, particularly in the light of her great affinity to digital technology. She joined management on October 1, 2007.

Steven Karger passed away on March 5, 2008 at the age of 48 years. He left behind a wife and three children. The publisher's death left a deep, painful gap in the company. Gabriella Karger now had to tackle the demanding task of his succession. In this way, rather than the original plan, she now continued the tradition of the family business as the representative of the fourth generation.

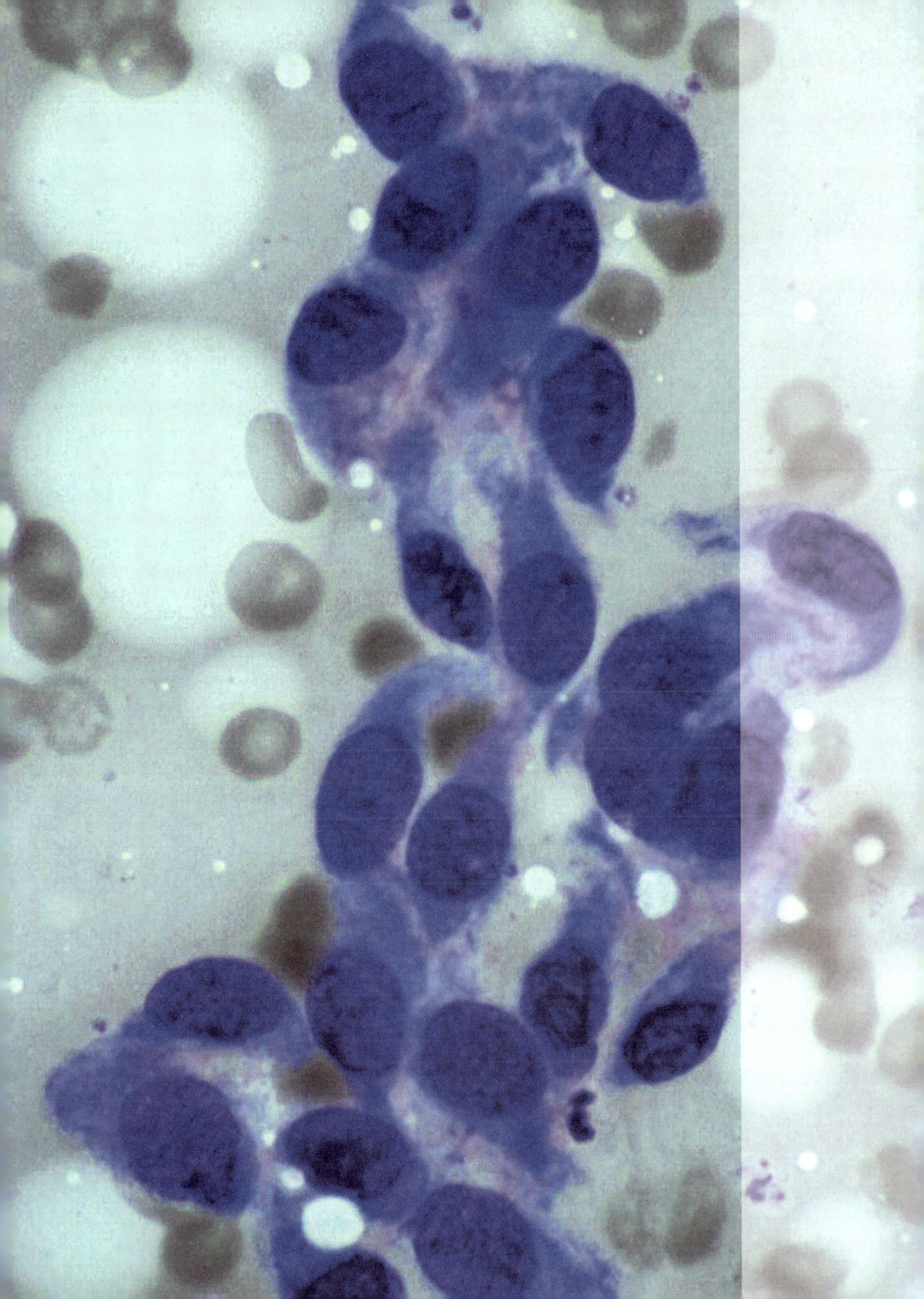

9

Team, Technologies, Strategies

KARGER

After the death of Steven Karger, the management and staff ensured that business continued. In this very difficult situation for the publishing house, Gabriella Karger and her team rallied. As president of the board of directors, Thomas Karger also remained in a key management position.

Like her brother, Gabriella Karger (CEO since 2012) considers it her central mission to anchor the publishing house in the digital world as an independent, family-run enterprise. And similarly to Steven Karger, who had begun to establish a team-oriented corporate culture, she is also keen to encourage a lively exchange of information and ideas. As a publisher, she considers herself part of a community in which all members have their specific tasks, but in which open communication is encouraged. She believes that it would be beneficial if everyone of the company had an insight into where problems and solutions may lie. The employees have therefore expressly been invited to work together towards the future of the publishing house, to think about new solutions and to join forces with the management to ensure that the company has an even more distinctive profile.

The collaborative approach of the management occasionally broke with familiar procedures and demanded that everyone think proactively. Gabriella Karger was and is, however, convinced that the publishing house will only continue to flourish if staff members can act with a degree of independence. She established a middle management level so that a larger group is involved in the internal transformation of the publishing house as well as in strategic planning, new products, services and sales channels.

The international financial crisis of 2008 also exacerbated the serials crisis. Like all academic publishers, after a delay of one to two years Karger suffered from the cutting of university budgets, especially in the United States. The tone of negotiations became harsher, and further price increases triggered uproar. Some institutions boycotted journals in protest against what they saw as exorbitant pricing policies, particularly among the large publishing houses. Karger had to assert its position more strongly in the competition for budgets in this heated atmosphere, but once again its personal and trusting relationship with customers and subscribers around the

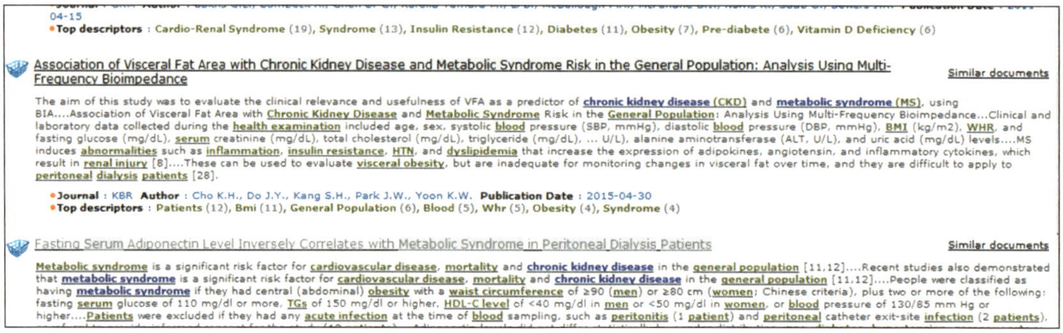

Advanced faceted searching of structured metadata allows topics and semantic relationships to be extracted and linked in the Karger content.

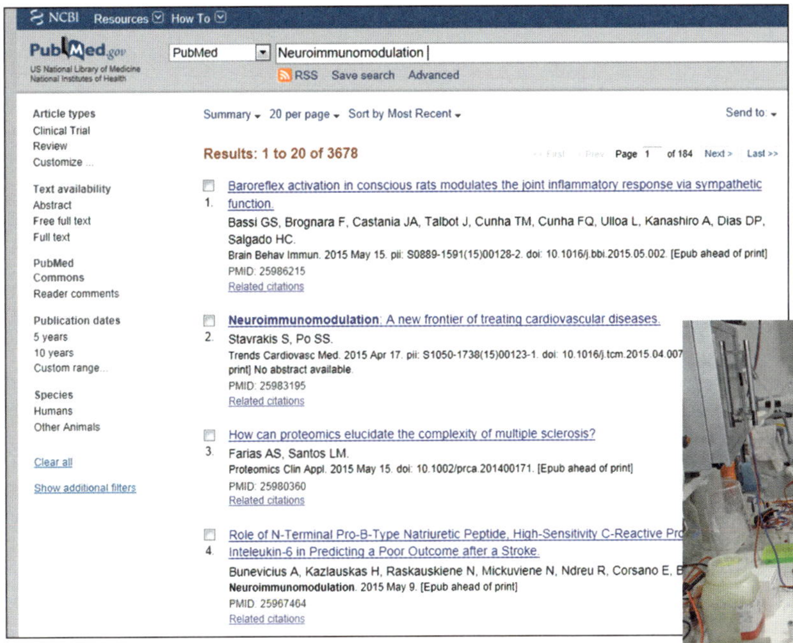

world paid off. Nonetheless, the discussion in the online business is increasingly dominated by usage and cost per download figures.

With its range of journals and books, Karger basically does not provide a different service than the big corporations; what distinguishes it is the more continuous and personal contacts it maintains with all stakeholders. Both the brokers and the users of information profit from this. Authors appreciate a careful, fast, and fair review process, personally supervised by the Editorial Office, and both authors and readers alike recognize the value of meticulously edited texts. This is why Karger continues to rely on its Copy Editing department that edits and proofreads the manuscripts accepted for publication – classic publishing services that have often already been abandoned by the large publishing houses to save costs. To Karger, clear and comprehensible texts, complete tables and captions as well as flawless lists of references, abbreviations, credits and high-quality illustrations make it easier to understand the content.

Being technically up-to-date has always been the standard for Karger. Whether IT solutions were developed in house or outsourced had to be decided case by case. Karger gradually departed from their long-held maxim of doing as much as possible on their own. Linking, connecting and establishing relationships between textual contents are of central importance. For example, articles published in Karger journals need to be easy to find both in the relevant scientific databases and via popular search engines. Therefore, like other publishers, Karger invested in software to semantically enrich content. This facilitates a more open search in which it is no longer necessary to name the object precisely. The advanced analysis no longer records only terms or character strings in the text, but relevant hits also result from a so-

‘Karger’s major strength lies in keeping an open mind and personal contacts to authors and editors, which remains a central task of the publishing house.’

Gabriella Karger (1964)

called semantic conceptual search. This enables a variety of possibilities for use: articles can be linked together and cross-referenced with additional literature. Around the year 2010, it seemed that the reception of scientific, and in particular medical, content in digital form was the standard. When it came to acquisitions, many libraries started making new publications available to their readers only in an electronic format. Digitalization made access and usage data quantifiable and quality aspects were decreasingly taken into consideration due to financial pressure. This particularly affected high-quality journals that were only in demand among specialized researchers working on very specific topics.

Prestigious journals with a high impact factor maintained their position, while others lost ground. Researchers searched for specific information and found it using modern search options. A single article from a journal, a chapter of a book or just a section of a publication might provide the required information. In terms of content sales, the publishing house had to adjust to smaller units. Its good experiences with sales of collections would be repeated with these smaller units. Karger tested the new business model in some 'in focus' compendia on topics such as diabetes or influenza. It had and continues to have promise, mainly because, as with the collections, it is possible to take advantage of the inter- and multidisciplinary product range. The focus is on the knowledge transfer for researchers in medicine, and the slogan remains 'Connecting the World of Biomedical Science'.

Karger's core business has been and still remains its now approximately 100 journals, 40 book series and other publications. The mainstay of its publishing activity is the intensive dialog with specialists within the medical scientific community, who, as experts in highly specialized subjects, also take responsibility for the content as editors of publications and are thus indispensable to the publishing house. The service the publisher renders consists of working closely with specialists to plan and supervise jour-

'Authors appreciate a careful, fast and fair review process, and Karger continues to rely on its Copy Editing department that edits and proofreads the manuscripts accepted for publication.'

nals and book series. Therefore, personal and regular contact continues to be one of the fundamental tasks of the publishing house.

Adopting an international perspective is crucially important when planning future developments, but this is nothing new. Since 1960, the English language has become firmly established in science and publishing as well as in Karger's offices in Basel and their branches elsewhere in the world. Meanwhile, it has also become important to adapt to the specific conditions of medicine, particularly in the fast-growing scientific and economic field in South America and Southeast Asia as well as in India and China.

The optimization of internal processes and workflows remains a high priority. Since publishing services are also offered by other providers, it is important to control and handle all publishing projects with great efficiency. Transparent cost calculations that can satisfy authors, information providers, and users are indispensable. Karger has a high standard of quality that can only be ensured in the future if the company operates profitably. Going forward, business will continue to revolve around the procurement and sale of editorial contents from the field of medicine. For its targeted moderate growth in the development of new markets, the publishing house sees potential both internationally and among new target groups on the procurement as well as the consumer side. A particular focus is on the pharmaceutical industry, for which Karger provides a wide range of products under the label of 'Pharma Services', ranging from contacts with editors to the joint development of publications all the way to the customized supply of editorial content for a wide variety of purposes. Karger continues to maintain close contacts focusing more than hitherto on academic medical societies that are relevant to the publishing program. Often, these are small, highly specialized societies, for which Karger is a suitable partner.

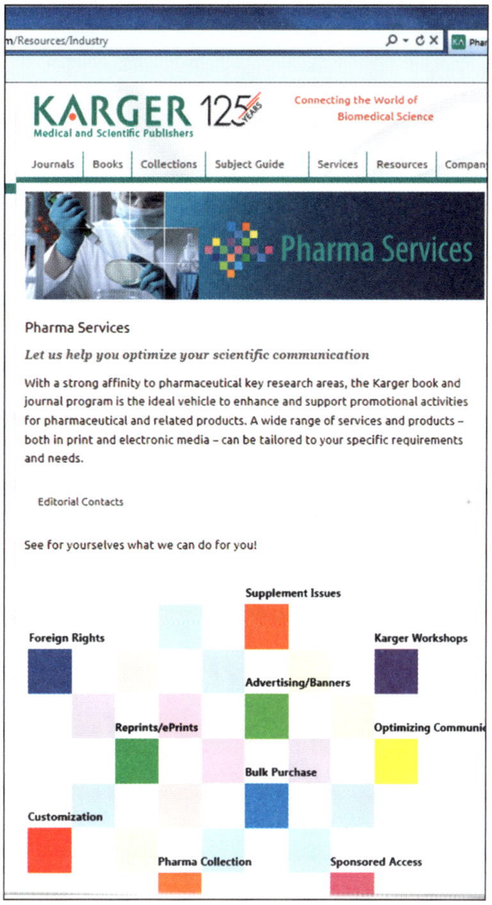

Providing knowledge about medicine will remain Karger's core business. Whilst in past decades a few disciplines enjoyed the limelight, attention nowadays tends to be focused on rare diseases and personalized medicine. In addition, translational medicine, which acts at the interface between basic research and clinical applications, and global public health are becoming increasingly important. Karger views the interdisciplinary approach that appears to be necessary as offering great potential and thus is

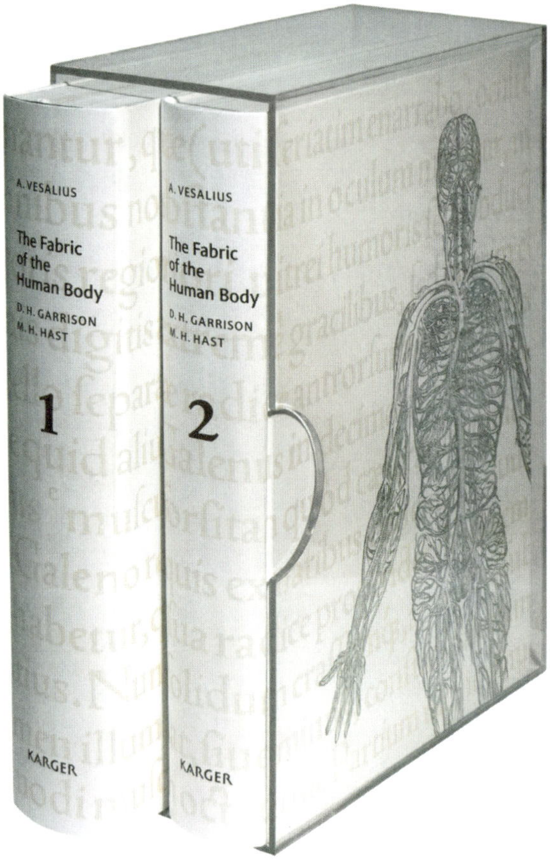

planning open access publications dealing with topics across various disciplines. New trends and topics are identified by means of the company's long-standing and well-maintained contacts with researchers around the world. Ideas for new projects and publications originate from the dialog with them. Karger's major strength lies in keeping an open mind with respect to the field of medicine and thus in continuously upgrading its program.

Karger's close relationship with medicine and researchers around the globe is combined with a passion for the tradition and the culture of the medical field. Between 2011 and 2013, the third generation of publishers, represented by Thomas Karger, and the fourth generation, represented by Gabriella Karger, had the opportunity to once again provide proof of this passion. With their team, they faced a very special challenge: the publication of the English translation of Andreas Vesalius' *De humani corporis fabrica*

(1543 and 1555), the Renaissance textbook that marked the beginning of modern anatomy. This challenge brings together everything that has distinguished Karger for 125 years and will continue to differentiate the publishing house. The Publisher's Note from the translation *The Fabric of the Human Body* (www.vesalius-fabrica.com) edited by Dan Garrison and Malcolm Hast (Chicago) summarizes it best:

> One fine day in 2009 Dan Garrison walked through the front door of Karger Publisher's headquarters in Basel, the town where the *Fabrica* was printed 466 years before, and proposed the publication of the first comprehensive and annotated translation of both versions of Andreas Vesalius' masterpiece in English, i.e. the 1543 and 1555 editions. At Karger we quickly realized that only a multinational cooperation could make such an ambitious project come true. And indeed, the combination of Karger's highly skilled Swiss craftsmanship with the considerable expertise of the US translators in the production of this 21st-century version of Vesalius' *Fabrica* was made possible by our main sponsor from Kuwait; thus, as was the case with the original, it became an international joint venture.
>
> In the 1540s, when the Flemish author Vesalius prepared the first edition of his epochal anatomical oeuvre in Padua, the city where he taught anatomy, he worked together with block-cutters in Venice. It was under the protection of Venetian and Milanese merchants that the 273 wooden, folio-sized plates found their way to Basel, Switzerland, where the humanist Erasmus and the scientist Paracelsus had also spread their knowledge.
>
> The challenge of creating a modern and user-friendly edition of the groundbreaking atlas by Andreas Vesalius was enormous. Even the original typeface was modernized; Christian Mengelt created a computer typeface from the original Basel Antiqua used in the 1543 edition. Now, over 5,500 footnotes and roughly 1,500 pages with an abundance of greatly enhanced illustrations are the final result, for which great credit is due to a highly motivated team.
>
> In the name of Karger Publishers we would like to thank each and every one both in front of and behind the scenes – our highly esteemed external partners as well as the Karger Vesalius team – who helped to make the New Fabrica into the fascinating experience which it has become.
>
> *Thomas Karger and Gabriella Karger*

It can't be predicted in what form medical information will be transferred in the future. While 25 years ago it seemed that printed products would not survive, today we observe that at least some publications continue to appear in print. Nonetheless, electronic media now dominate the publishing industry. It remains to be seen whether the open access model will prevail or what hybrid forms will emerge. In addition, it will be decisive to always stay up-to-date in indexing and linking articles. Today, people expect that information is immediately available worldwide even if doubts sometimes arise within the scientific community and among publishers of whether greater speed in itself has any value.

Founder Samuel Karger's maxim was 'Allzeit voran' ('Always forward'), to which Gabriella Karger adds 'Gemeinsam allzeit voran' ('Always forward together'). It is only with its dedicated employees that the publishing house will be able to remain the successful, internationally active company that it is and make its motto 'Connecting the World of Biomedical Science' a reality.

Connecting the World of Biomedical Science

KARGER 125 YEARS

Medical and Scientific Publishers

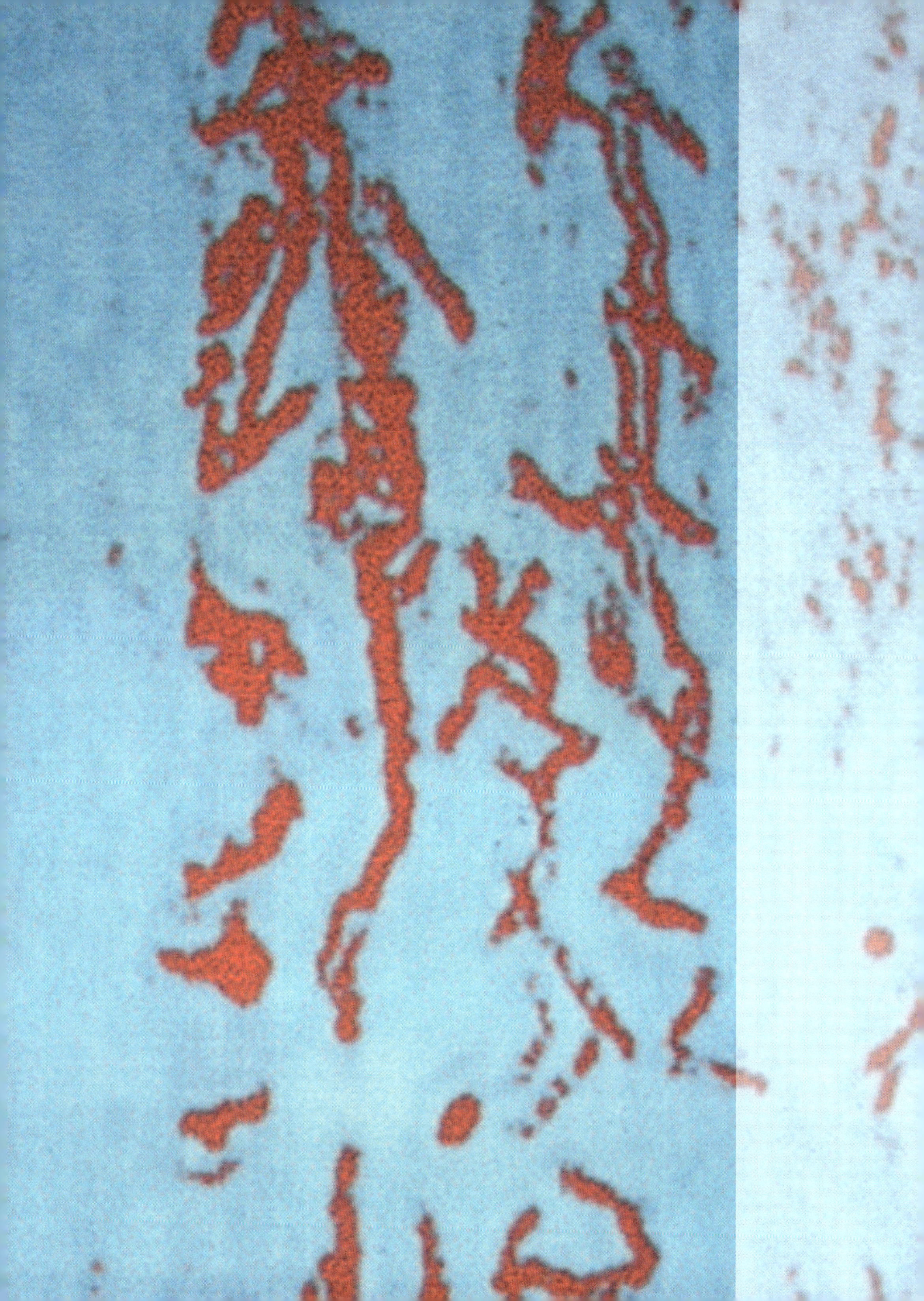

Epilogue

The Medical Publisher in the Information Age

Kent R. Anderson

The shift from print to online as the main information delivery platform for medical journals has made the role of the medical publisher more important than ever, with more media types, more media outlets, more article-based publishing, and more content types. Authors demand effective and coordinated release of their studies, readers require more levels of review and data sharing, and the technology itself requires high levels of experimentation on complex infrastructure.

With the worldwide scope of many studies and author groups, medical publishers are at the epicenter of activities that are more intricate than ever before. Large, general medical journals and smaller specialist journals are equally affected.

In a rather counterintuitive response to these new and emergent demands, some observers have predicted 'the end of publishers' and an era of open communication unencumbered by organized information businesses and editorial gatekeepers.

Nevertheless, publishers have become more important, especially in scholarly medical publishing, where quality information is most urgently needed, peer-reviewed articles and journals must compete effectively with less rigorously reviewed alternative information sources, and a focus on the growing body of specialized knowledge is at the forefront of a shared mission between researchers, practitioners, and publishers.

While the primary roles of the medical publisher – organizing editorial functions and peer review, launching and maintaining journals, and managing the legal, financial, and technological functions – are well known, this article highlights some of the less obvious but increasingly important roles of the medical publisher.

To start, we need to define why publishers exist.

What Is a Publisher?

Clay Shirky famously said of the Internet Age [1]:

> Publishing is not evolving. Publishing is going away. Because the word 'publishing' means a cadre of professionals who are taking on the incredible difficulty and complexity and expense of making something public. That's not a *job* anymore. That's a *button*. There's a button that says 'publish', and when you press it, it's done.

This attempt to reduce publishing to 'a button' reveals a very poor understanding of what a publisher does. Publishing entails far more than the final functional steps of making works public. In this particular context, Shirky's examples consist of individuals posting on platforms like WordPress, Blogger, Facebook, and Twitter. He misunderstands the roles even for these companies, but in an enlightening manner.

Like any other publisher, these newer technology companies have had to build infrastructure and invest in new business development for long periods before gaining prominence and achieving sustainable income levels. While their business models differ from some publishing business models, the notion of producing free information on a widespread basis in order to generate revenue indirectly through advertising is nothing new – many controlled circulation publications have done that for decades. Many still do, from the local arts papers given out freely at kiosks to even medical journals. The Internet has changed some aspects of the model – the scale, the pace, and the targeting of associated advertising – but the fundamental bargain is preserved. These new publishers also face legal, commercial, and brand risks, which they manage in various ways.

Publishers take risks on behalf of authors and readers in order to make information available.

In this light, a more accurate picture of publishers emerges – publishers take risks on behalf of authors and readers in order to make information available.

This perspective also clarifies another confusion Shirky introduces – that is, if companies like WordPress and Twitter are publishers, then the individuals using them to reach readers are authors. Therefore, 'authorship' is the button Shirky is seeing. Publishing is a much more complex job.

Because these platforms make it seem that authorship is granted with the press of a button, Shirky and others assume that authors on these new publishing platforms are unencumbered by gatekeepers. In fact, editorial controls are written into the platforms, from character limits (Twitter) to copyright protections (YouTube) and various controls on allowable words (Facebook, Twitter). The editor is an algorithm, but as with any publisher, there is a gatekeeper. In addition, authors must register, provide a verified way to be contacted (e-mail address and/or mobile number), and agree to lengthy terms of use.

A publisher takes risks in order to make information available, and attempts to minimize the risk and attract better readers through editorial, legal, and marketing controls. This is a vital definition to set out early, so that the increasing importance of medical publishers in the Information Age can be clearly seen.

Filters, Curation, and Marketing

Shirky does, however, correctly diagnose one aspect of the Internet Age – the concept of 'filter failure', i.e. the way users can be overwhelmed, misdirected, underserved, or misled by ineffective selection processes. The concept of filter failure is very important to medical publishers, as they have unique pressures on the quality of information they publish – there is much more at stake with publications that potentially affect the health and well-being of men, women, and children.

'Filtering' is a term currently in vogue to describe editorial selection processes. Most medical journals accept approximately 1 in 5 articles submitted, with some large, general medical journals accepting as few as 1 in 20 or 25 submissions. This selection process helps to define the journal and often the field, as top-tier general and specialist journal articles indicate to some extent where research success is occurring and what editors are seeking and find interesting.

'Curation' is a term borrowed from museums to describe the process of arranging relevant information so that users get a broader, sequential, and deeper view of a topic.

Both filtering and curation have changed significantly as the Internet has developed. In the print world, editors would place similar articles one after the other in bound, printed issues. They could use textual or design cues to point readers to other material in the same issue. The thinking was guided by the form – each bound issue had clear boundaries, a beginning, an end, and an order that emerged from these constraints.

Search engines have broken these bonds. Issues are no longer the defining form of scholarly communication. Like all other fields, medicine is now based on an article economy. This has allowed more article-based publishing practices to emerge. These approaches have proven important during public health crises, as editors have been able to focus on particular important articles and rush them online irrespective of issue constraints. Over the past decade, these capabilities have helped medical experts and practitioners deal with SARS, H1N1, anthrax, Ebola, and other emerging diseases and issues in health care.

Because search engines drive traffic directly to articles, along with the continued effectiveness of e-mail notifications, journal home pages are of diminishing importance. More traffic goes directly from search engines to article pages, bypassing curation attempts. This has increased the complexity of marketing published content, especially as the number of outlets interested in new journal publications has expanded and fragmented.

There are three major implications to these changes in the publications themselves.

First, articles must carry more weight in the filtering and curation process. Consequently, article designs now contain more boxes, more 'take home' points, and more pointers to related content. Article landing pages on journal sites have more links to related content and design elements that seek to bring curation to the article level.

Second, publishers must understand and adapt to third-party curation tools like Google and Bing, which operate by their own rules. These rules change frequently, and some changes have caused entire journals to vanish from major search indices for a time. Complying with these rules is a new requirement to filtration and curation.

Finally, curation is now a much more technical matter than ever before, with metadata (data surrounding content sources) and technical rules guiding traffic, blocking certain nefarious elements, and making content more relevant to searchers.

Publishers and Trust, Relevance, and Communities

When a field establishes a journal, the domain passes into an important stage of development. Publishers have traditionally been partners in crossing this threshold. In fact, as publishers bear much of the risk in development around a journal, their investments in editors, editorial boards, marketing, and infrastructure are integral to new communities of practice developing. Journals also give the field a stamp of authority that other activities – meetings, seminars, and scattered research in other journals – cannot.

Over time, this authority and the publisher's investment will create a community that can be trusted, in both the large sense (permanence) and the local sense (trust in the practitioners).

In addition to bringing a community together, publishers help to sense when a new community might be developing or branching off. In my days working in pediatrics, the field of neonatology was reaching a critical point in its scientific and professional advancement. New techniques and approaches for the care for premature infants in imaging, in genetics and genomics, and in microsurgery had allowed neonatologists to develop a specialized community to care for newborns. We invested to help pull together disparate case reports, collated large image archives that could assist with diagnoses, and started new journals for the field. These activities elevated the intellectual work of the field while also giving it added legitimacy.

In similar ways and in prior eras, publishers used their natural contacts and professional training to become integral to developing areas such as critical care medicine, emergency medicine, family medicine, orthopedic surgery, genetics, and tropical medicine. Currently, what I call 'field detection' is developing content in translational medicine, hospitalist medicine, and advanced nursing fields. As international and regional variations become prominent, this function has become more important and complex.

There is no way to know how many problems medical journals head off through their enforcement of editorial rigor and transparency guidelines and policies.

Enforcing Transparency

One way that journals engender trust is by creating more transparency around the financial ties and possible motivations of authors. This is very important in medical publishing, and requires complex policies and thoughtful implementation of these policies and guidelines. It also requires a commercial enterprise that is motivated to create and preserve this type of transparency. Publishers take on this role.

As biomedical research and the stakes involved in the pharmaceutical and medical technology industries have all increased, disclosure rules and conflict of interest guidelines have become more important to maintaining and improving trust in the medical literature. Allegations of bias – for positive outcomes, for commercially viable outcomes – have been a consistent feature of medical publishing for decades. There have been some egregious cases in which financial interests clearly led to data manipulation, data suppression, and bias.

These trends are not receding. In fact, they seem to be growing in intensity, as the Internet makes global business more lucrative and lowers barriers to entry for those who wish to make medical claims.

There is no way to know how many problems medical journals head off through their enforcement of editorial rigor and transparency guidelines and policies, but my experience with a number of specialty journals and major general medical journals leads me to believe that journals and publishers prevent a major meltdown of the core of biomedical research. If every claim, conflict, and bias were allowed to go through unchallenged and uncorrected, medical research would suffer its version of Chernobyl, and it would take decades to recover.

In most cases, these preventative measures occur outside the light of day. However, there are times when items slip through, and the potential for damage can be great. Publishers bring the legal discipline and resources to bear on these matters, backing up editors, maintaining insurance, and hiring sharp copy editors who often are the ones to identify problems. Ultimately, publishers are the gatekeepers and decide what gets published and what does not.

Teaching Editors, Creating Intellectual Nexus, and Maintaining Continuity

Not all researchers and medical doctors are natural editors, and they are often surprised and somewhat overwhelmed by what a single year of editorial work involves. The training, reassurance, management continuity, and succession planning involved in strong editorial products throughout medical publishing are largely due to publishers and their commitment to editorial outputs.

Granted, because publishers have taken large risks in developing their businesses, they have a clear interest in minimizing editorial risks. This entails having strong, prominent, intelligent, seasoned, and well-trained editors at all levels. A natural and beneficial side-effect of this is that good publishers are able to offer stronger editorial products on a more consistent basis. Even if editors retire or new publications are launched, the publisher's editorial infrastructure is strong enough to absorb the changes or growth initiatives.

Early speculation about disruption of traditional journals and publishers failed to contemplate the consequences. Imagine if within a 2- to 3-year period your most important specialty journal simply vanished, for no good reason. Worse, imagine if it and all of its counterparts vanished in the same time frame. These brands are vital signals of quality and relevance. Many medical journals are now more than 100 years old, and the continuity these provide is important not only to the historical record of each field, but also to the aspirations of trainees and young researchers, who can aim to be published in these venerated but vital journals.

Journals also create an intellectual nexus in each field in which knowledgeable researchers and practitioners face questions of research integrity, quality, relevance, and bias every day. At the highest levels, this work can profoundly change a field, as new trends in research analysis or ethics are sorted out by editors over the course of a number of early papers or submissions. Publishers tacitly support these hothouses of ideas, partly out of practical concerns about how strong editorial offices reduce risk for the publisher, but also for many out of their own sense of duty and idealism. Medical publishers are, in my view, among the most idealistic people you are likely to meet, and at minimum appreciate the subtleties and importance of such activities. The best actively participate in a constructive and supportive manner.

Letting Scientists Do Science, Letting Practitioners Practice

Looking at scholarly medical publishers as service providers is another way to gain a useful perspective on their overall importance to science and scholarship, as well as why their role will continue to grow as the information landscape becomes more complex and global.

Scientists and medical practitioners are specialists. As with all specialists, there are points where they hire other specialists to tackle tasks that are either tangential to their main work or which require special skills and knowledge. Publishers fulfill both of these roles, in addition to eliminating risk for authors who can effectively delegate duties like editing, design, branding, marketing, legal protection, and archival support.

One of the major risks and roles publishers adopt is the creation, management, and maintenance of peer review infrastructure, panels, and systems. A hidden cost of scholarly and academic publishing is the cost of rejection. Every paper submitted to a journal creates expenses, which the 2012 PEER study estimated at USD 250 per paper [2]. These estimates have proven out empirically, but are often misinterpreted. These are the expenses associated with handling a submission just to initial decision, which in most cases is limited to 'send out to peer review or not'. Most journals have a level of decision making – an 'editorial rejection' or 'desk rejection' – that involves editors determining a paper is not suitable for quality, novelty, or relevance reasons. Once a paper goes beyond this point, more costs accrue as more staff, editorial, and system time is used.

One of the major risks and roles publishers adopt is the creation, management, and maintenance of peer review infrastructure, panels, and systems.

Medical journals have some of the lowest acceptance rates, with most around 20% and some of the more selective journals in the 4–5% range. This means that those few papers selected for publication have to support the costs, staff, and effort involved in rejecting 80–96% of the total manuscripts submitted, in addition to supporting their full costs of publication, providing for management and basic corporate duties (human resources, legal, building, and insurances), and providing a sufficient margin for the business to be self-sufficient and worth pursuing.

Peer review systems are not trivial systems to build, maintain, and manage. Reviewers have to be solicited, monitored, and assessed regularly. Reviewers often number in the hundreds if not thousands, and categorizing these people, monitoring communications, maintaining deadlines, and so forth, is a major effort.

The seemingly simple task of getting objective reviews done for papers is something that would require significant amounts of time and attention from physicians, researchers, and other practitioners. In addition, because each would be a non-specialist, their efforts would be less efficient, increasing the waste in the system.

Besides the practical activities that save researchers and practitioners time and effort, there is another service being provided: objective, third-party assessment together with a strong partner that protects the final paper from abuse and misuse.

A medical research report issued from the university in which the senior author resides would generate some attention, but as has been shown again and again, the general field would not have a reliable signal about the paper's possible merits in this case, and may discount the university's publication of the work as merely intramural support of the author. In addition, after the grant has been expended, the university in most cases would have no incentive to continue to protect the work, migrate it to new technologies, or monitor it for abuses, while a publisher has long-term interests aligned with activities that protect and maintain the published work.

These are subtle but important aspects of the value medical publishers add to the community, to the work of researchers, and to the productivity of researchers and practitioners.

Conclusion

Recently, I published the third annual update to a post on the blog I founded, The Scholarly Kitchen. Originally entitled, '60 Things Publishers Do', three years later the list has grown into '82 Things Publishers Do' [3]. The original post was in response to claims that publishers only enlist volunteer peer reviewers and do light copyediting before charging the producers of the content once again to read it.

Publishers are an important part of the academic and scholarly system. They are a part of the community, providing services and taking on duties others cannot or will not do. These duties include organizing and managing blinded peer-review systems, taking risks on behalf of authors today and tomorrow, building enduring brands that can be trusted, developing and evolving editorial workflows and expertise, and much more.

In today's world, when there is more information than ever, when time seems to be scarcer, and when trusting what you read is more important than in the past, everything publishers do seems more important than ever – every risk they take, every girder of continuity they contribute, every community they help develop, and every high mark of success they represent.

References

1 http://genius.com/Clay-shirky-how-we-will-read-annotated.
2 http://www.peerproject.eu/introductory-final-presentations/.
3 http://scholarlykitchen.sspnet.org/2014/10/21/updated-80-things-publishers-do-2014-edition/.

Kent R. Anderson is the Publisher of *Science* at the AAAS. Previously, he worked as CEO/Publisher of the STRIATUS/JBJS, Inc., a publishing executive at the *New England Journal of Medicine*, and Director of Medical Journals at the American Academy of Pediatrics.

The unabridged version of this article appears in the 2015 edition of the *Karger Gazette* (Anderson KR: The medical publisher in the Information Age: the digital era means more media, more technology, and more important roles. *Karger Gazette* 2015;74:2–4, www.karger.com/gazette).

Selected Publications*

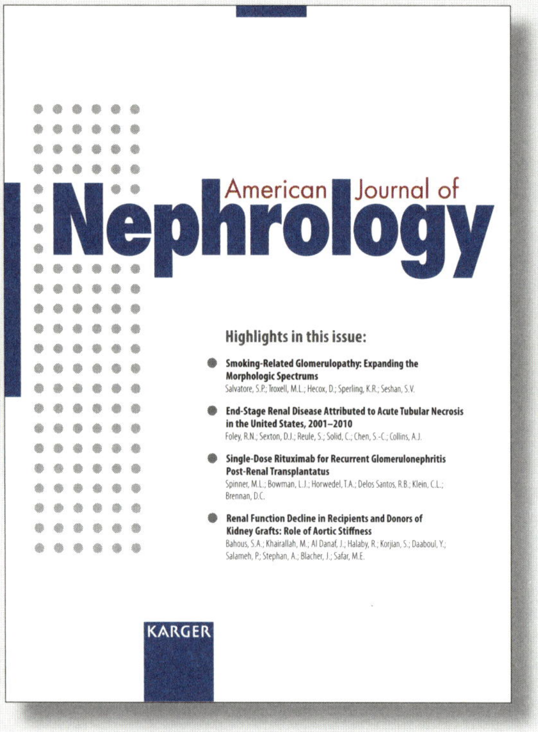

American Journal of Nephrology

The *American Journal of Nephrology* is a peer-reviewed journal that focuses on timely topics in both basic science and clinical research. Papers are divided into several sections, including:

- Original reports, spanning clinical, basic and outcomes research
- In-depth topic reviews on specific topics selected by the Editor and Associate Editors
- Invited debates – selected on 'hot topics' determined by the editorial board

Responding to the ever increasing pace in research, the *American Journal of Nephrology* adheres to a very tight publication schedule: accepted papers appear online within 5–6 weeks of acceptance.

American Journal of Nephrology
Citation: Am J Nephrol
ISSN: 0250–8095
e-ISSN: 1421–9670
www.karger.com/AJN

*For the complete list of Karger journals, book series and individual book titles see www.karger.com.

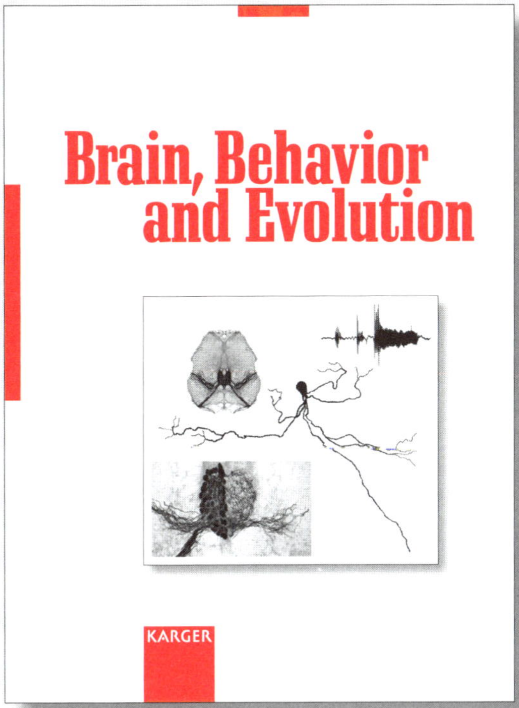

Brain, Behavior and Evolution

Editor-in-Chief
Georg F. Striedter, Irvine, Calif.

Brain, Behavior and Evolution is a journal with a loyal following, high standards, and a unique profile as the main outlet for the continuing scientific discourse on nervous system evolution. The journal publishes comparative neurobiological studies that focus on nervous system structure, function, or development in vertebrates as well as invertebrates. Approaches range from the molecular over the anatomical and physiological to the behavioral. Despite this diversity, most papers published in *Brain, Behavior and Evolution* include an evolutionary angle, at least in the discussion, and focus on neural mechanisms or phenomena. Some purely behavioral research may be within the journal's scope, but the suitability of such manuscripts will be assessed on a case-by-case basis. The journal also publishes review articles that provide critical overviews of current topics in evolutionary neurobiology.

Brain, Behavior and Evolution
Citation: Brain Behav Evol
ISSN: 0006-8977
e-ISSN: 1421-9743
www.karger.com/BBE

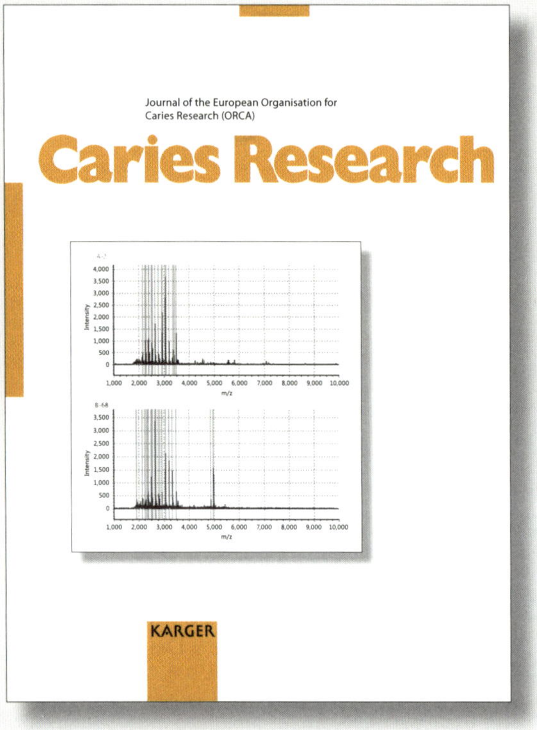

Caries Research

Caries Research publishes epidemiological, clinical and laboratory studies in dental caries, erosion and related dental diseases. Some studies build on the considerable advances already made in caries prevention, e.g. through fluoride application. Some aim to improve understanding of the increasingly important problem of dental erosion and the associated tooth wear process. Others monitor the changing pattern of caries in different populations, explore improved methods of diagnosis or evaluate methods of prevention or treatment. The broad coverage of current research has given the journal an international reputation as an indispensable source for both basic scientists and clinicians engaged in understanding, investigating and preventing dental disease.

Caries Research
Citation: Caries Res
ISSN: 0008-6568
e-ISSN: 1421-976X
www.karger.com/CRE

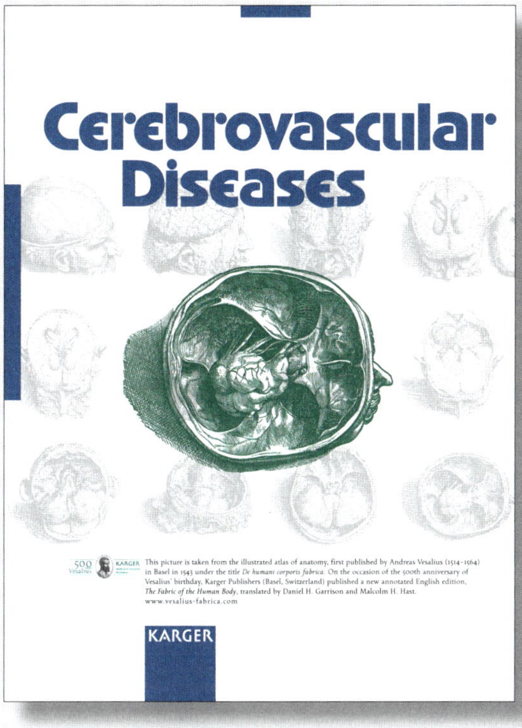

Cerebrovascular Diseases

Editor-in-Chief
Michael G. Hennerici, Mannheim, Germany

Associate Editors
Harold P. Adams Jr., Iowa City, USA
Craig Anderson, Camperdown, Australia
Dale Corbett, Ottawa, Canada
Larry B. Goldstein, Durham, USA
Lorenz Hirt, Lausanne, Switzerland
Jong-Sung Kim, Seoul, Republic of Korea
Yoshihiro Kokubo, Osaka, Japan
Stephen Meairs, Mannheim, Germany
Man Mohan Mehndiratta, New Delhi, India
Jay P. Mohr, New York, USA
Yukito Shinohara, Tokyo, Japan
Narayanaswamy Venketasubramanian, Singapore, Singapore

A rapidly-growing field, stroke and cerebrovascular research is unique in that it involves a variety of specialties such as neurology, internal medicine, surgery, radiology, epidemiology, cardiology, hematology, psychology and rehabilitation. *Cerebrovascular Diseases* is an international forum which meets the growing need for sophisticated, up-to-date scientific information on clinical data, diagnostic testing, and therapeutic issues, dealing with all aspects of stroke and cerebrovascular diseases. It contains original contributions, reviews of selected topics and clinical investigative studies, recent meeting reports and work-in-progress as well as discussions on controversial issues. All aspects related to clinical advances are considered, while purely experimental work appears if directly relevant to clinical issues.

Cerebrovascular Diseases
Citation: Cerebrovasc Dis
ISSN: 1015–9770
e-ISSN: 1421–9786
www.karger.com/CED

Dementia
and Geriatric Cognitive Disorders

Editor-in-Chief
V. Chan-Palay, Singapore

As a unique forum devoted exclusively to the study of cognitive dysfunction, *Dementia and Geriatric Cognitive Disorders* concentrates on Alzheimer's and Parkinson's disease, Huntington's chorea and other neurodegenerative diseases. The journal draws from diverse related research disciplines such as psychogeriatrics, neuropsychology, clinical neurology, morphology, physiology, genetic molecular biology, pathology, biochemistry, immunology, pharmacology and pharmaceutics. Strong emphasis is placed on the publication of research findings from animal studies which are complemented by clinical and therapeutic experience to give an overall appreciation of the field.

Dementia and Geriatric Cognitive Disorders
Citation: Dement Geriatr Cogn Disord
ISSN: 1420–8008
e-ISSN: 1421–9824
www.karger.com/DEM

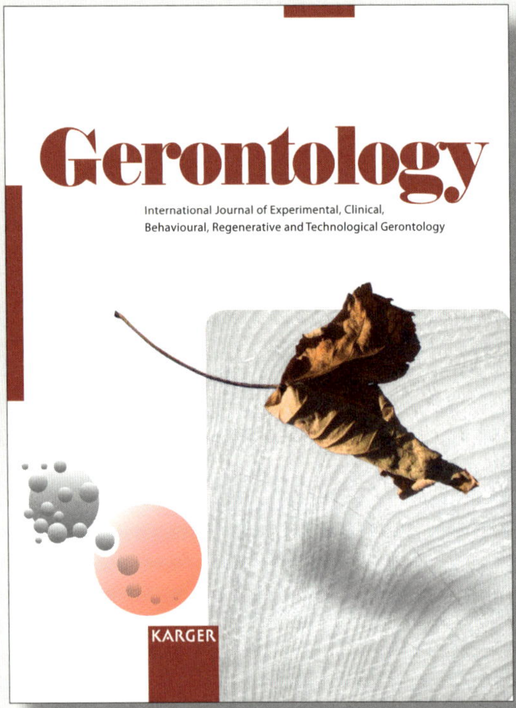

Gerontology

Editor-in-Chief
G. Wick, Innsbruck

Editors
A. Bartke, Springfield, Ill.
M.E. Gershwin, Davis, Calif.
D. Gerstorf, Berlin
S. Gravenstein, Cleveland, Ohio
J.D. Henry, St. Lucia, Qld.
M. Lechleitner, Zirl
G. Lepperdinger, Salzburg (Regeneration)
B. Najafi, Tucson, Ariz. (Technology)
J. Vijg, Bronx, N.Y.

In view of the ever-increasing fraction of elderly people gerontology has become the major scientific discipline that may provide clues to successfully deal with this substantial personal, medical and sociological problem. *Gerontology*, the oldest journal in the field, is the only one that publishes gerontological research from a very broad perspective, rather than focusing on single issues of aging research. The range of papers is classified into four major sections. In each section, Original Papers, Mini-Reviews, Viewpoints and critical Debates are published and will be complemented by short systematic Reviews.

In the Clinical Section, the aetiology, pathogenesis, prevention and treatment of age-related diseases are discussed from a gerontological rather than a geriatric viewpoint. The Experimental Section contains up-to-date contributions from basic gerontological research. Papers dealing with behavioural development and related topics are placed in the Behavioural Science Section. Basic aspects of regeneration in different experimental biological systems as well as in the context of medical applications are dealt with in a special section that also contains information on technological advances for the elderly.

Providing a primary source of high-quality papers covering all aspects of aging in humans and animals, *Gerontology* serves as an ideal information tool for all readers interested in the topic of aging from a broad perspective.

Gerontology
Citation: Gerontology
ISSN: 0304–324X
e-ISSN: 1423–0003
www.karger.com/GER

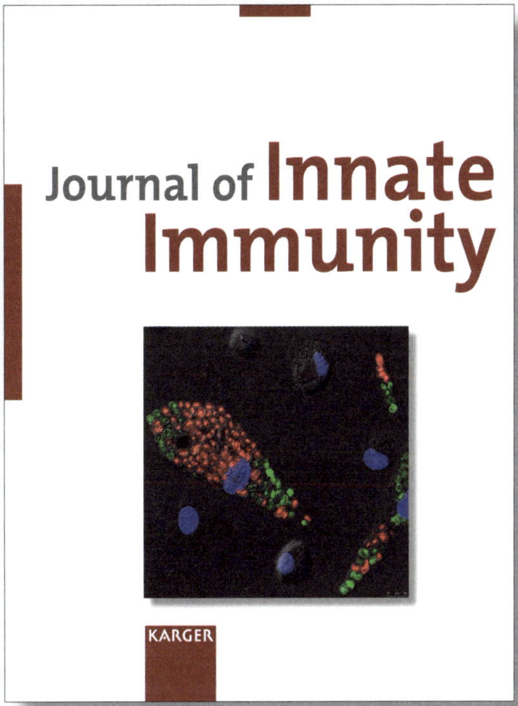

Journal of Innate Immunity

Editors-in-Chief
H. Herwald, Lund
A. Egesten, Lund

The *Journal of Innate Immunity* is a bimonthly journal covering all aspects in the area of innate immunity, including evolution of the immune system, molecular biology of cells involved in innate immunity, pattern recognition and signals of 'danger', microbial corruption, host response and inflammation, mucosal immunity, complement and coagulation, sepsis and septic shock, molecular genomics, and development of immunotherapies.

The journal publishes original research articles, short communications, reviews, commentaries and letters to the editors. In addition to regular papers, some issues feature a special section with a thematic focus.

Journal of Innate Immunity
Citation: J Innate Immun
ISSN: 1662–811X
e-ISSN: 1662–8128
www.karger.com/JIN

Neuro-endocrinology

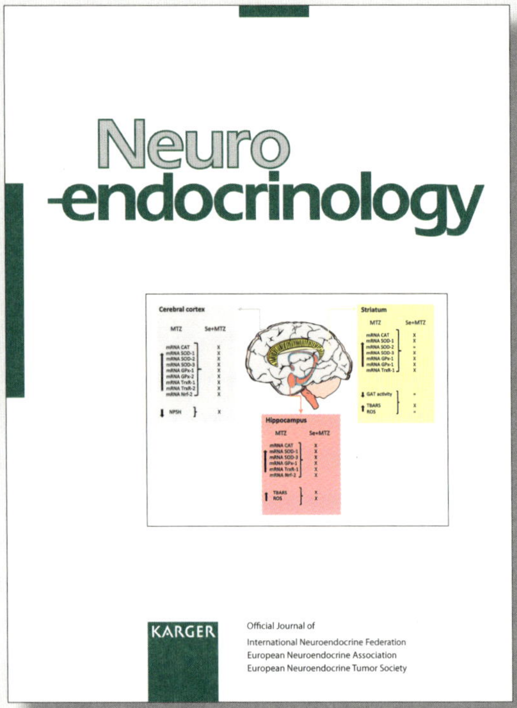

Neuroendocrinology publishes papers reporting original research in basic and clinical neuroendocrinology. The journal explores the complex interactions between neuronal networks and endocrine glands (in some instances also immunecells) in both central and peripheral nervous systems. Original contributions cover all aspects of the field, from molecular and cellular neuroendocrinology, physiology, pharmacology, and the neuroanatomy of neuroendocrine systems to neuroendocrine correlates of behaviour, clinical neuroendocrinology and neuroendocrine cancers. Readers also benefit from reviews by noted experts, which highlight especially active areas of current research, and special focus editions of topical interest.

Neuroendocrinology
Citation: Neuroendocrinology
ISSN: 0028-3835
e-ISSN: 1423-0194
www.karger.com/NEN

Oncology Research and Treatment

With the first issue in 2014, the journal *Onkologie* has changed its title to *Oncology Research and Treatment*. By this change, publisher and editor set the scene for the further development of this interdisciplinary journal. The English title makes it clear that the articles are published in English – a logical step for the journal, which is listed in all relevant international databases. For excellent manuscripts, a 'Fast Track' was introduced: The review is carried out within 2 weeks; after acceptance the papers are published online within 14 days and immediately released as 'Editor's Choice' to provide the authors with maximum visibility of their results. Interesting case reports are published in the section 'Novel Insights from Clinical Practice' which clearly highlights the scientific advances which the report presents.

Oncology Research and Treatment is the official organ of the Deutsche Gesellschaft für Hämatologie und Medizinische Onkologie e.V. (DGHO), the Arbeitsgemeinschaft Internistische Onkologie in der deutschen Krebsgesellschaft e.V. (AIO) and the Deutsche Fatigue Gesellschaft e.V. (DFaG). Bridging disciplines, the monthly issues inform at a high scientific level about clinical trials as well as oncology training and clinical issues related to all questions of cancer research and treatment. The title change and re-orientation did not alter the scope and frequency of publication. The terms and conditions of subscription also remain unchanged.

Oncology Research and Treatment
Citation: Oncol Res Treat
ISSN: 2296–5270
e-ISSN: 2296–5262
www.karger.com/ORT

Psychotherapy and Psychosomatics

Editor-in-Chief
G.A. Fava, Bologna

Associate Editors
F. Cosci, Firenze
E. Tomba, Bologna

As the volume of literature in the fields of psychotherapy and psychosomatics continues to grow, it becomes increasingly difficult to keep abreast of new and important developments, and to have independent, critical views on key clinical issues concerned with assessment and treatment of mental disorders. *Psychotherapy and Psychosomatics* has progressively increased its reputation for independence, originality and methodological rigor. It has anticipated and developed new lines of research concerned with psychosomatic medicine, psychotherapy research and psychopharmacology. The journal features editorials and reviews on current and controversial issue; innovations in assessment and treatment; case management articles; original investigations; literature updates; and a lively correspondence section. It has become a unique and vital reference to current research and practice, the home of innovative thinking at the interface between medical and behavioral sciences.

Psychotherapy and Psychosomatics
Citation: Psychother Psychosom
ISSN: 0033-3190
e-ISSN: 1423-0348
www.karger.com/PPS

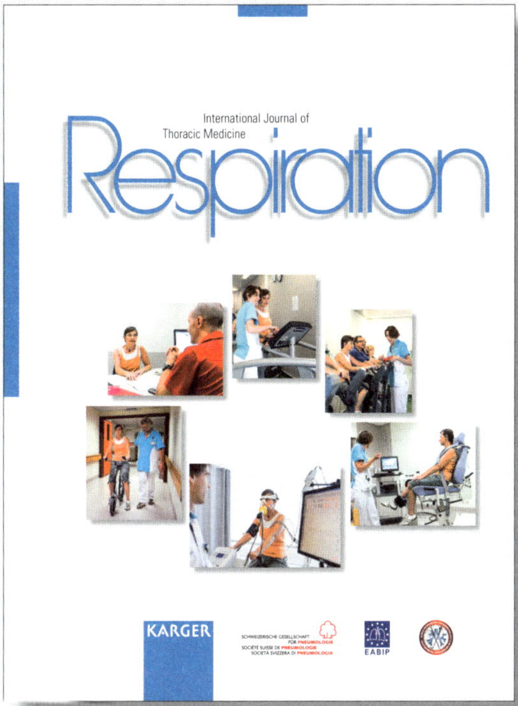

Respiration

Respiration brings together the results of both clinical and experimental investigations on all aspects of the respiratory system in health and disease. Clinical improvements in the diagnosis and treatment of chest and lung diseases are covered, as are the latest findings in physiology, biochemistry, pathology, immunology and pharmacology. The journal includes classic features such as editorials that accompany original articles in clinical and basic science research, reviews and letters to the editor. Further sections are: Technical Notes, The Eye Catcher, What's Your Diagnosis?, The Opinion Corner, New Drugs in Respiratory Medicine, New Insights from Clinical Practice and Guidelines. *Respiration* is the official journal of the Swiss Society for Pneumology (SGP) and also home to the European Association for Bronchology and Interventional Pulmonology (EABIP), which occupies a dedicated section on Interventional Pulmonology in the journal. This modern mix of different features and a stringent peer-review process by a dedicated editorial board make *Respiration* a complete guide to progress in thoracic medicine.

Respiration
Citation: Respiration
ISSN: 0025–7931
e-ISSN: 1423–0356
www.karger.com/RES

Chemical Immunology and Allergy

Editor: *J. Ring*, München

Appearing for the first time in 1939, *Progress in Allergy* was the first and, for many years, the only international review series in the English language devoted to immunology. The present title, *Chemical Immunology and Allergy*, reflects the rapidly increasing role of concepts and techniques of molecular and cell biology, as well as the greater emphasis now being placed on research into the biologic problems of immunology, hypersensitivity and immunopathology. Highly specialized, single-topic volumes keep both investigators and practicing physicians up to date with the latest developments in areas of immunology and allergy where recent research has been concentrated.

ISSN: 1660–2242
e-ISSN: 1662–2898
www.karger.com/CHIAL

Contributions to Nephrology

Editor: C. *Ronco*, Vicenza

The speed of developments in nephrology has been fueled by the promise that new findings may improve the care of patients suffering from renal disease. Participating in these rapid advances, this series has released an exceptional number of volumes that explore problems of immediate importance for clinical nephrology. Focus ranges from discussion of innovative treatment strategies to critical evaluations of investigative methodology. The value of regularly consolidating the newest findings and theories is enhanced through the inclusion of extensive bibliographies which make each volume a reference work deserving careful study.

ISSN: 0302–5144
e-ISSN: 1662–2782
www.karger.com/CONEP

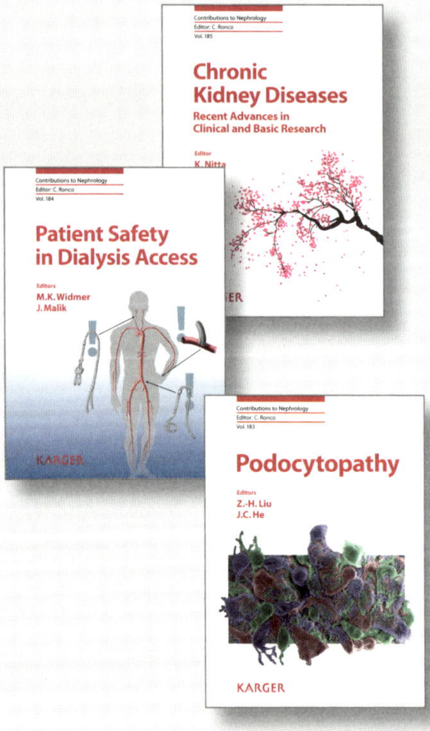

Current Problems in Dermatology

Editors: *P. Itin*, Basel; *G.B.E. Jemec*, Roskilde

This series presents a selection of new ideas and information in the field of clinical and investigative dermatology. Crossing the borders of skin diseases it explores interdisciplinary strategies, linking dermatology with immunology, allergology and other research areas. Each volume is individually designed to provide an up-to-date presentation of new and interesting topics in dermatology and skin research. The contributing authors are acknowledged experts in their fields. The wide range of topics discussed includes normal and pathological skin conditions, the development of new equipment, drugs and analytical techniques, and recent advances in the treatments that can be offered to patients.

ISSN: 1421–5721
e-ISSN: 1662–2944
www.karger.com/CUPDE

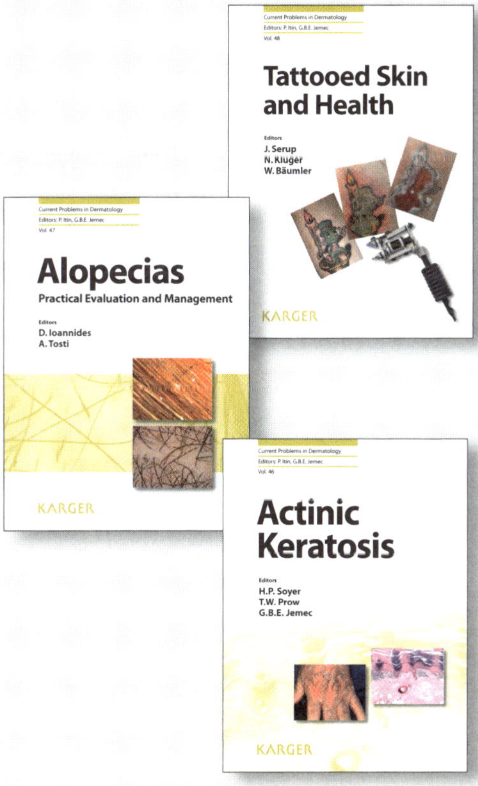

Developments in Ophthalmology

Editor: *F. Bandello*, Milan

The book series *Developments in Ophthalmology* provides ophthalmologists, ophthalmic surgeons and practicing optometrists with expert reviews highlighting the most important developments in their field. Innovative procedures from a wide range of areas – such as corneal or macular disorders, pharmacological issues, eye surgery, eye banking and many more – are presented with the appropriate technical information required to encourage rapid introduction into clinical practice. For each volume expert authors producing articles of only the highest standards are invited to present their knowledge.

During the considerable history of its existence *Developments in Ophthalmology* has established itself as an essential reference in the field of ophthalmology.

ISSN: 0250–3751
e-ISSN: 1662–2790
www.karger.com/DEOPH

Endocrine Development

Editor: *P.-E. Mullis*, Bern

This series is devoted to specific areas of fetal, neonatal, pediatric and adolescent endocrinology. It addresses a wide range of relevant issues in the context of a well defined subject and covers new areas of clinical and basic research. Clear evidence-based guidelines on clinical management by highly qualified basic and clinical scientists who present their clinically relevant recent findings are also provided. Bringing together experts from related disciplines such as fetal and perinatal medicine, epidemiology, public health, molecular endocrinology, radiology and adult endocrinology, the volumes cover a broad spectrum in endocrine development.

ISSN: 1421–7082
e-ISSN: 1662–2979
www.karger.com/ENDEV

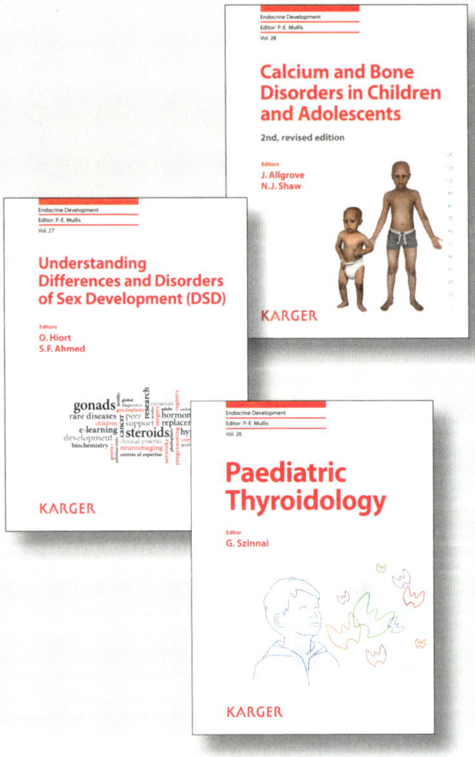

Frontiers of Hormone Research

Editors: *E. Ghigo*, Turin; *F. Guaraldi*, Turin; *A. Benso*, Turin

New sophisticated technologies and methodological approaches in diagnostics and therapeutics have led to significant improvements in identifying and characterizing an increasing number of medical conditions, which is particularly true for all aspects of endocrine and metabolic dysfunctions. Novel insights in endocrine physiology and pathophysiology allow for new perspectives in clinical management and thus lead to the development of molecular, personalized treatments. In view of this, the active interplay between basic scientists and clinicians has become fundamental, both to provide patients with the most appropriate care and to advance future research.

The individual volumes of this series explore cutting-edge topics in the field of endocrinology and metabolism, providing the most updated, critical opinions of international leading researchers and clinicians. Going from in vitro studies to daily clinical applications, each volume presents a state-of-the-art overview including a discussion of future perspectives.

ISSN: 0301–3073
e-ISSN: 1662–3762
www.karger.com/FHORE

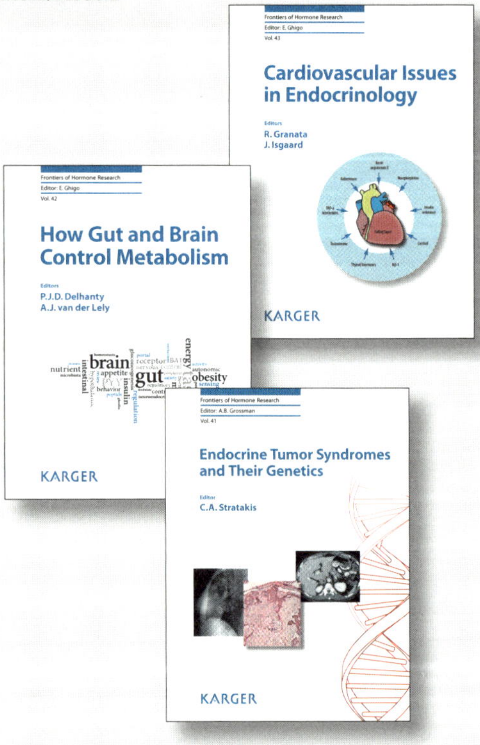

Frontiers of Neurology and Neuroscience

Editor: *J. Bogousslavsky,* Montreux

Focusing on topics in the fields of both Neurosciences and Neurology, this series provides current and unique information in basic and clinical advances on the nervous system and its disorders.

ISSN: 1660–4431
e-ISSN: 1662–2804
www.karger.com/FNENE

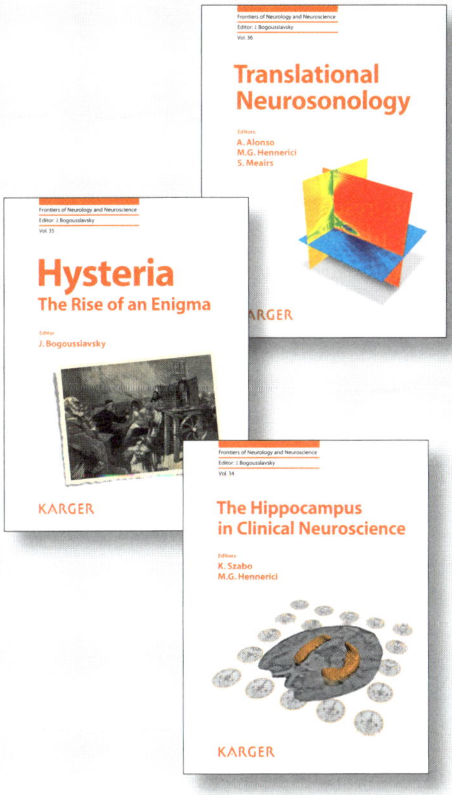

Monographs in Oral Science

Editors: *A. Lussi,* Bern;
M.C.D.N.J.M. Huysmans, Nijmegen;
H.-P. Weber, Boston, Mass.

For over three decades, *Monographs in Oral Science* has provided a source of in-depth discussion of selected topics in the sciences related to stomatology. Senior investigators are invited to present expanded contributions in their fields of special expertise. The topics chosen are those which have generated a long-standing interest, and on which new conceptual insights or innovative biotechnology are making considerable impact. Authors are selected on the basis of having made lasting contributions to their chosen field and their willingness to share their findings with others.

ISSN: 0077–0892
e-ISSN: 1662–3843
www.karger.com/MOORS

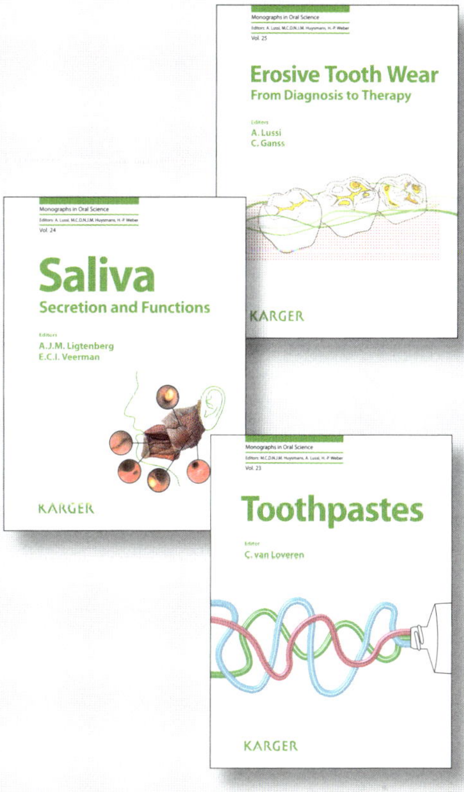

Nestlé Nutrition Institute Workshop Series

In 1981, *Nestlé Nutrition* was created with the aim of improving and disseminating knowledge on health and nutrition, as well as providing scientific information to the medical profession. Initially, *Nestlé Nutrition* consisted of the book series *Nestlé Nutrition Workshop Series: Pediatric Program*, which since 1997 has been extended to include also the *Nestlé Nutrition Workshop Series: Clinical & Performance Program*. In 2011, these two book series merged to form the *Nestlé Nutrition Institute Workshop Series*. One of the major components of the series is the organization of workshops and the publication of the proceedings. Each workshop focuses on the latest findings, the controversies and further research needs concerning various topics in nutrition. The books of this series will be of great assistance to pediatricians, clinical investigators, sports scientists as well as health workers and nutritionists concerned with both adult and pediatric health and nutrition.

ISSN: 1664–2147
e-ISSN: 1664–2155
www.karger.com/NNIWS

Progress in Respiratory Research

Editor: *F.J.F. Herth*, Heidelberg

Important aspects of the respiratory system, from its role in the alimentation and protection of the organism to its susceptibility to disease, receive attention in the monographs published in this series. Some volumes present debates on topics vital to respiratory research. Others represent the unique experiences of a single renowned author and stand as first-time treatments of specific issues. The series is designed for researchers as well as clinicians and extends to them a collection of valuable reference tools.

ISSN: 1422–2140
e-ISSN: 1662–3932
www.karger.com/PRRER

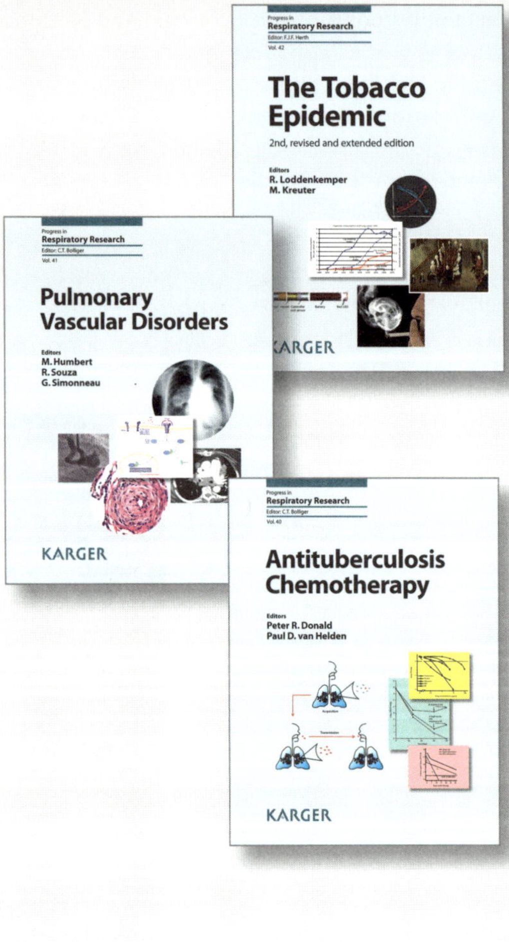

Atlas of Cytoarchitectonics of the Adult Human Cerebral Cortex

1st English ed.

Authors: *C. von Economo*, Vienna; *G.N. Koskinas*, Athens; Editor: *L.C. Triarhou*, Thessaloniki

A milestone in neuroscience research, this high-profile Atlas depicts the cellular structure of practically every area of the human cortex with direct applications to current research in brain function. The entirety of the 112 original microphotographic plates, brilliant achievements in scientific microphotography and representing the 107 cytoarchitectonic areas of the human cerebral cortex, are reproduced in full size – large enough to be used for teaching purposes. An extensive introduction places the cytoarchitectonic studies of von Economo and Koskinas in a historical as well as a modern perspective, summarizing the essence of their findings and providing Brodmann area correlations. Biographies of von Economo and Koskinas and complete listings of their hard-to-find works are included in the Appendix.

Originally published in German in 1925, it was considered a 'royal gift to science'. Revising Brodmann's nomenclature of 1909, the Nobel prize nominee von Economo and his colleague Koskinas took cytoarchitectonics to a new zenith, filling in gaps left by Brodmann on normal cortical structure, and documenting detailed findings in the frontal, parietal, temporal and occipital lobes, the insula, hippocampus, and superior limbic region.

Far from being of purely academic or historical interest, this essential guide for all research on the cerebral cortex is of fundamental value to investigators in the brain and behavioral sciences, including basic, cognitive and evolutionary neuroscience, neuroanatomy, neurophysiology, neuroimaging, neuropsychology and neurolinguistics, as well as to physicians in the clinical fields of neurology, neuropathology, neurosurgery and psychiatry.

Atlas of Cytoarchitectonics of the Adult Human Cerebral Cortex
Translated, revised and edited with an Introduction and additional appendix material by L.C. Triarhou (Thessaloniki)
With full-scale reproductions of the original 112 microphotographic plates, including 8 tables, 33 figures, 4 in color. In acrylic cassette
Authors: von Economo C. (Vienna), Koskinas G.N. (Athens)
Editor: Triarhou L.C. (Thessaloniki)
1st English ed.
X + 182 p., hard cover, 2008
ISBN: 978-3-8055-8289-6

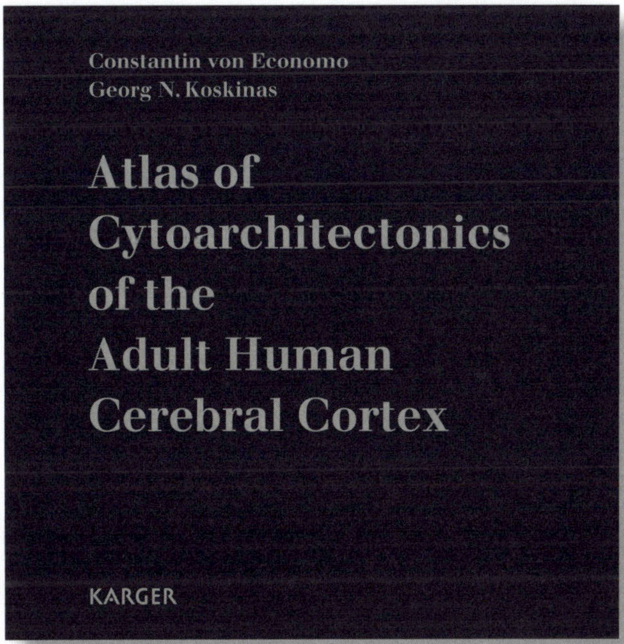

BMA
BMA Book Award
'highly commended' in Neurology

Biofunctional Textiles and the Skin

Editors: *U.-C. Hipler*, Jena; *P. Elsner*, Jena

In recent years the development of new technologies has permitted the production of 'functional' or 'smart' textiles. These fabrics are capable of sensing changes in environmental conditions or body functions and are adequately responding to them. They are able to absorb substances from the skin or to release therapeutic or cosmetic compounds. For instance, they can be used in underwear with an integrated cardio-online system or as textiles with carrier molecules. The focal point of interest in biofunctional textiles lies currently on the use of textiles supporting therapy and prevention in dermatology.

This volume collects information about new trends in the interaction between textiles and the skin, particularly the development of antimicrobial finished textiles. It presents a selection of papers which will contribute to further consolidate the dialogue between dermatologists, allergologists, biomaterial scientists and textile engineers.

Current Problems in Dermatology, Vol. 33
Biofunctional Textiles and the Skin
Editors: Hipler U.-C. (Jena), Elsner P. (Jena)
X + 204 p., 53 fig., 7 in color, 29 tab., hard cover, 2006
ISBN: 978–3–8055–8121–9
e-ISBN: 978–3–318–01349–8

BMA Book Award
'commended' in Dermatology

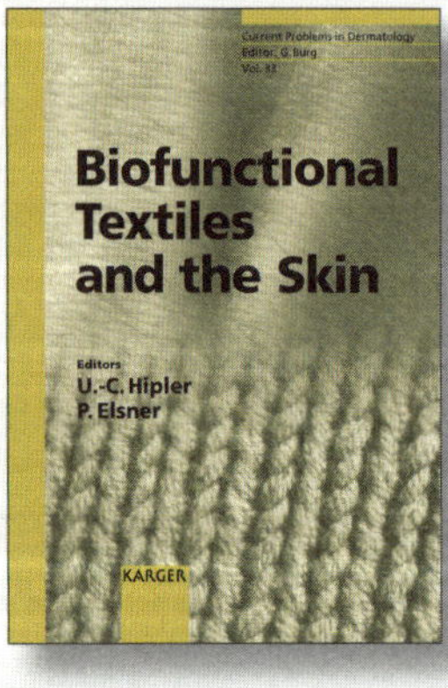

Diagnostics of Endocrine Function in Children and Adolescents

4th, revised and extended edition

Editors: *M.B. Ranke*, Tübingen; *P.-E. Mullis*, Bern

A multitude of new developments, not only in the rapidly advancing field of molecular genetics and steroid metabolism but in all traditional areas of pediatric endocrinology, have influenced the diagnostic approach in children and adolescents with endocrine disorders, thus warranting this 4th, revised and extended edition of *Diagnostics of Endocrine Function in Children and Adolescents*.

Several chapters have been revised completely and all have been thoroughly updated. In addition, new chapters dealing with the muscle-bone unit and bone metabolism have also been incorporated. The original format of the chapters, which are a combination of in-depth discussion of the diagnostic process, practical conclusions and expert advice based on extensive experience, was maintained. Easy-to-use tables and figures allow for quick reference. Flowcharts of possible diagnostic pathways lead to the most frequent diagnoses.

Presenting a broad range of diagnostic approaches, test procedures, and normative data required for establishing diagnoses for a broad spectrum of endocrine disorders, this book is an indispensable reference tool not only for endocrinologists and pediatricians but also for professionals in other specialties seeking evidence-based, rapid diagnostic solutions as the basis of advice and therapy for their patients.

Diagnostics of Endocrine Function in Children and Adolescents
Editors: Ranke M.B. (Tübingen), Mullis P.-E. (Bern)
4th, revised and extended edition
XII + 536 p., 109 fig., 80 tab., hard cover, 2011
ISBN: 978–3–8055–9414–1
e-ISBN: 978–3–8055–9415–8

Diffuse Parenchymal Lung Disease

Editors: *U. Costabel*, Essen; *R.M. du Bois*, London; *J.J. Egan*, Dublin

Diffuse parenchymal lung disease (DPLD) represents a large and heterogeneous group of disorders. Although new insights into the pathogenesis and new techniques such as high-resolution-CT have led to a better understanding of DPLD, clinical management remains a challenge.

This volume is designed to become a valuable aid in the diagnosis and management of DPLD. It starts with general topics such as a new classification system for DPLD, the general diagnostic approach, and clinical evaluation of the patient by radiology, histological patterns and bronchoalveolar lavage. This is followed by consideration of some basic aspects such as the genetics of DPLD, the principles of granuloma formation, pulmonary fibrogenesis and vasculitis, as well as novel treatment approaches. A large part of the book consists of disease-specific chapters, which discuss granulomatous disorders, idiopathic pulmonary fibrosis and other entities of idiopathic interstitial pneumonia, the collagen vascular diseases, drug-induced infiltrative lung disease as well as orphan diffuse lung diseases including Langerhans' cell histiocytosis, lymphangioleiomyomatosis and pulmonary alveolar proteinosis. Special chapters on DPLD in children and lung transplantation for end-stage fibrosis complete this comprehensive publication.

Written by internationally known experts, this well-organized volume is recommended to a wide audience including respiratory physicians in training, practice and research as well as radiologists and pulmonary pathologists.

Progress in Respiratory Research, Vol. 36
Diffuse Parenchymal Lung Disease
Editors: Costabel U. (Essen),
du Bois R.M. (London),
Egan J.J. (Dublin)
X + 348 p., 144 fig., 15 in color,
47 tab., hard cover, 2007
ISBN: 978-3-8055-8153-0
e-ISBN: 978-3-318-01377-1

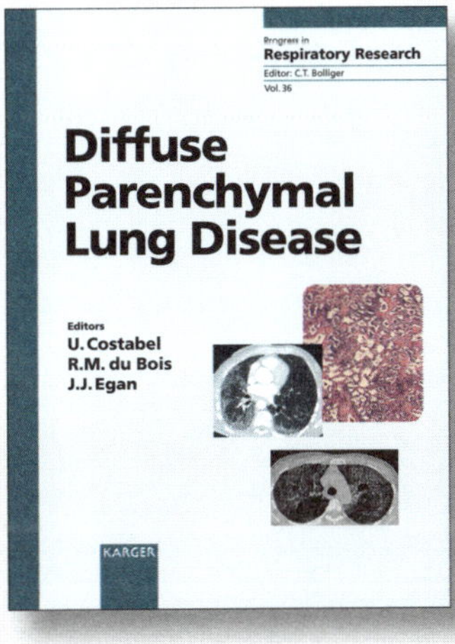

BMA

BMA Book Award
'highly commended'
in Respiratory Medicine

Drug Hypersensitivity

Editor: *W.J. Pichler*, Bern

Drug hypersensitivities are immune-mediated reactions characterized by exanthems, fever and internal organ involvement. They represent a challenging field of research: many different drugs can induce a variety of clinical symptoms through multiple mechanisms of drug-host interaction, many of which are poorly understood.

This is the first book to approach the phenomenon of drug hypersensitivity in a comprehensive manner. Besides epidemiological aspects, it addresses the immunological mechanisms underlying these complicated reactions which go far beyond the IgE-mediated drug allergies also considered in the book. Furthermore, the book covers clinical manifestations and new diagnostic methods, and introduces some recently established animal models. Well-accepted and also completely new concepts are presented and discussed in detail. Many topics are treated from multiple perspectives, and the 33 chapters are thoroughly cross-referenced.

This publication will be of immense value not only to allergologists, dermatologists and anyone prescribing medication, but also to scientists in a pharmaceutical industry challenged by the economic effects of failures in drug development or drug removal from the market. Elucidating the mechanisms of drug hypersensitivity will not only help to identify patients at risk but will also provide novel insights into the pathophysiology of numerous immune-mediated diseases.

Drug Hypersensitivity
Editor: Pichler W.J. (Bern)
X + 438 p., 78 fig., 58 in color, 78 tab., hard cover, 2007
ISBN: 978-3-8055-8269-8
e-ISBN: 978-3-318-01454-9

BMA

BMA Book Award
'commended'
in Basic and Clinical Sciences

ISCN 2013

An International System for Human Cytogenetic Nomenclature (2013)

Editors: *L.G. Shaffer*, Spokane, Wash.; *J. McGowan-Jordan*, Ottawa, Ont.; *M. Schmid*, Würzburg

This publication extends the now classic system of human cytogenetic nomenclature prepared by an expert committee and published in collaboration with *Cytogenetic and Genome Research* since 1963. Revised and finalized by the ISCN Committee and its advisors at a meeting in Seattle, Wash., in April 2012, the ISCN 2013 updates, revises and incorporates all previous human cytogenetic nomenclature recommendations into one systematically organized publication that supersedes all previous ISCN recommendations.

There are several new features in ISCN 2013:

- an update of the microarray nomenclature, many more illustrative examples of uses of nomenclature in all sections
- some definitions including chromothripsis and duplication
- a new chapter for nomenclature that can be used for any region-specific assay.

The ISCN 2013 is an indispensable reference volume for human cytogeneticists, technicians and students for the interpretation and communication of human cytogenetic nomenclature.

ISCN 2013
An International System for Human Cytogenetic Nomenclature (2013)
Recommendations of the International Standing Committee on Human Cytogenetic Nomenclature Published in collaboration with *Cytogenetic and Genome Research*
Plus fold-out: 'The Normal Human Karyotype G- and R-bands'
Editors: Shaffer L.G. (Spokane, Wash.), McGowan-Jordan J. (Ottawa, Ont.), Schmid M. (Würzburg)
VI + 140 p., 11 fig., 4 tab., soft cover, 2013
ISBN: 978-3-318-02253-7

Macular Edema

Editors: *G. Coscas*, Créteil, Paris; *J. Cunha-Vaz*, Coimbra; *A. Loewenstein*, Tel Aviv; *G. Soubrane*, Créteil, Paris

For many years diagnosing macular edema at an early stage has been a main difficulty. Recently, multiple and remarkable advances of modern imaging technologies which allow recognition of the main etiologies, have immensely improved its diagnosis and treatment.

In this volume different patterns and etiologies of macular edema are described and the importance of preserving the photoreceptors at an early stage in order to retain center visual acuity is discussed. It brings together the most recent data and evidence-based medicine, and, on the other hand, the pathophysiological basis of macular edema and the different approaches for drug delivery to the posterior segment are presented. Recommendations for treatment procedures or different therapies have been carefully analyzed and considered prior to inclusion.

Macular Edema – A Practical Approach provides the ophthalmologist with a synthesis of knowledge to diagnose, to determine etiology, and to offer the patient viable treatment options for his benefit.

Developments in Ophthalmology, Vol. 47
Macular Edema
A Practical Approach
Editors: Coscas G. (Créteil, Paris), Cunha-Vaz J. (Coimbra) Loewenstein A. (Tel Aviv), Soubrane G. (Créteil, Paris)
XII + 204 p., 115 fig., 77 in color, 8 tab., hard cover, 2010
ISBN: 978-3-8055-9434-9
e-ISBN: 978-3-8055-9435-6

BMA

BMA Book Award
'highly commended'
in Surgical Specialities

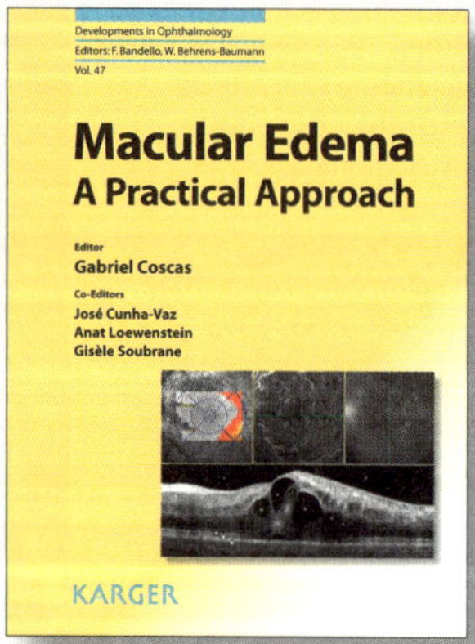

Molecules of Life and Mutations

Understanding Diseases by Understanding Proteins

Author: *S. Schwarz,* Innsbruck

This book provides insights into the structures and functions of 130 of the most important biomolecules and their interactions with other endogenous or exogenous molecules. These interactions are illustrated by 3-dimensional images of their atomic structures rather than by abstract formulas or acronyms. The author has compiled an extraordinary collection of molecules which he has visualized in pictures of stunning clarity and beauty by applying molecular modelling software to their atomic coordinate files (deposited in the Brookhaven Protein Data Bank, PDB). Together with short explanatory texts they provide the reader with a deepened understanding of biological phenomena in the normal as well as the diseased organism.

Following a novel didactic approach, this book distills the most important facts in a clear and concise manner, thus making it accessible to the nonspecialist reader as well; at the same time it serves as a guide to the vast amount of information stored in authoritative Internet databases such as the PDB and OMIM (Online Mendelian Inheritance in Men) databanks. For an in-depth study of structures and genetic diseases, an enhanced index listing databank accession numbers, gene name and chromosomal location of each molecule under consideration is added, which can also be accessed online on the book's website. A detailed list of useful Internet resources rounds off the practical value of this textbook.

Molecules of Life and Mutations
Understanding Diseases by Understanding Proteins
Author: Schwarz S. (Innsbruck)
XIV + 144 p., 475 fig., 475 in color, hard cover, 2002
ISBN: 978-3-8055-7395-5

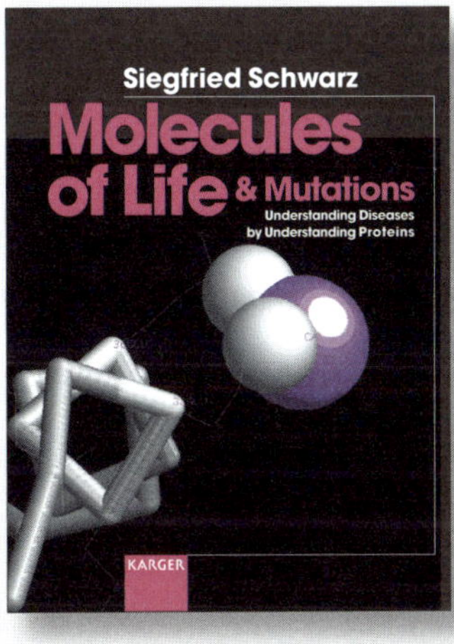

BMA

BMA Book Award
'highly commended' in
Basic and Clinical Sciences

Olszewski and Baxter's Cytoarchitecture of the Human Brainstem

3rd, revised and extended edition

Editors: *J.A. Büttner-Ennever,* Munich; *A.K.E. Horn,* Munich

The introduction serves as a concise, up-to-date treatise on the role of reticular formation, explaining both historical and current concepts of its function.

Uniquely combines both low- and high-powered photographic documentation of the complete human brainstem, whereby low-magnification plates show the overall organization and name each individual region or nucleus, while the accompanying high-magnification photos document in superb detail the different features of each individual nucleus which characterize and identify it.

As the only atlas of the human brainstem available containing full descriptions of organization, function and connectivity of more than 100 individual nuclei, it also comprises state-of-the-art summaries of the current knowledge on each nucleus including a comprehensive references list; it has almost become a lexicon of the human brainstem. Other aspects of the brainstem such as effects of lesions and immunochemical properties are also covered.

Olszewski and Baxter's
Cytoarchitecture of the Human Brainstem
Editors: Büttner-Ennever J.A. (Munich), Horn A.K.E. (Munich)
3rd, revised and extended edition
XII + 290 p., 604 fig., 1 tab., hard cover, 2014
ISBN: 978-3-318-02367-1

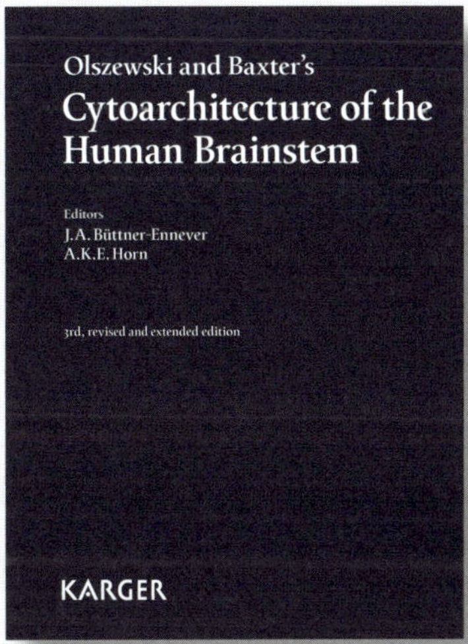

BMA

BMA Book Award
'highly commended'
in Neurology

Paediatric Bronchoscopy

Editors: *K.N. Priftis*, Athens;
M.B. Anthracopoulos, Patras; *E. Eber*, Graz;
A.C. Koumbourlis, Washington, D.C.;
R.E. Wood, Cincinnati, Ohio

This book provides an in-depth 'state-of-the-art' critical review of the technique and the applications of flexible and rigid bronchoscopy to infants and children. Written by an international panel of experts, it reviews the technical aspects of the procedure, its common and highly specialized applications as well as its potential alternatives. The chapters are enhanced by a wealth of original figures. A major innovation of the book is the inclusion of online videos from actual bronchoscopies that further illustrate and elaborate on the information provided in the text. The book is addressed to paediatric pulmonologists and otorhinolaryngologists, both experienced and in training, as well as to other personnel who are involved in the performance of the procedures. In addition, it is recommended to pulmonologists for adults and general paediatricians who need to be able to evaluate the usefulness of bronchoscopy for their patients and be aware of its limitations and potential contra-indications.

Progress in Respiratory Research, Vol. 38
Paediatric Bronchoscopy
Editors: Priftis K.N. (Athens), Anthracopoulos M.B. (Patras),
Eber E. (Graz), Koumbourlis A.C. (Washington, D.C.),
Wood R.E. (Cincinnati, Ohio)
X + 212 p., 136 fig., 86 in color, 37 tab., hard cover
+ online supplement material, 2010
ISBN: 978-3-8055-9310-6
e-ISBN: 978-3-8055-9311-3

BMA

BMA Book Award
'highly commended'
in Paediatrics

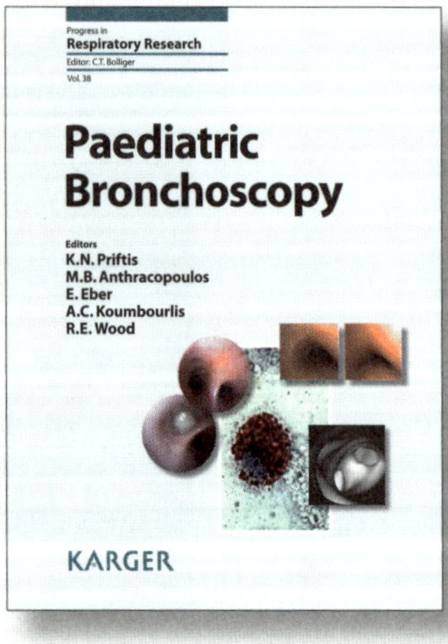

Practical Algorithms in Pediatric Endocrinology

(Practical Algorithms in Pediatrics.
Series Editor: *Z. Hochberg*)

2nd, revised edition

Editor: *Z. Hochberg*, Haifa

Algorithms provide a logical, concise and cost-effective approach to medical reasoning: utilizing a concise, step-by-step approach based upon clues from the history, physical examination and laboratory studies, algorithms help avoid excessive unnecessary procedures and testing.

The 2nd, revised edition of *Practical Algorithms in Pediatric Endocrinology* deals with practical issues of child growth, puberty, diseases of the endocrine glands, sexual differentiation, as well as aberrations of water, electrolyte, mineral and carbohydrate metabolism. Fifty clinical issues are covered by an algorithmic approach, breaking down long lists and tables of differential diagnosis into smaller, more manageable ones. Common clinical symptoms, signs and laboratory abnormalities are classified as they present themselves at the patient's bedside.

This book is aimed at general practitioners and pediatricians, in particular those who are not exposed to pediatric endocrine problems on a daily basis, and at trainees in endocrinology and pediatric endocrinology as they acquire familiarity with clinical problem solving to make rational choices when facing clinical dilemmas.

Practical Algorithms in
Pediatric Endocrinology
(Practical Algorithms in Pediatrics. Series Editor: Z. Hochberg)
Editor: Hochberg Z. (Haifa)
2nd, revised edition
IV + 112 p., 60 fig., 3 tab., spiral binding, 2007
ISBN: 978-3-8055-8220-9
e-ISBN: 978-3-8055-8375-6

Prader-Willi Syndrome

Coping with the Disease – Living with Those Involved

Author: *U. Eiholzer*, Zürich

Over the past years, research into the Prader-Willi syndrome (PWS) and its treatment options has progressed enormously and new insights have greatly improved the quality of life of those affected.

For many years, Urs Eiholzer has been engaged in the care of children with PWS and their families. He has published a great number of seminal scientific articles and contributions and supports a comprehensive, integrated treatment approach, taking into account both the biological and the psychological aspects of the syndrome. This easily understandable book is specifically suited for non-professionals and presents the clinical picture of the syndrome, its genetic cause, state-of-the-art research results and treatment options as used in the author's out-patient clinic. The text is illustrated with graphics and complemented by an abundance of pictures.

This publication addresses not only the family members of the affected children, but also those who have contact with them – from the neighbor to the pediatrician.

Prader-Willi Syndrome
Coping with the Disease – Living with Those Involved
Author: Eiholzer U. (Zürich)
120 p., 68 fig., 57 in color, soft cover, 2005
ISBN: 978-3-8055-7846-2

BMA
BMA Book Award
'highly commended'
in Popular Medicine

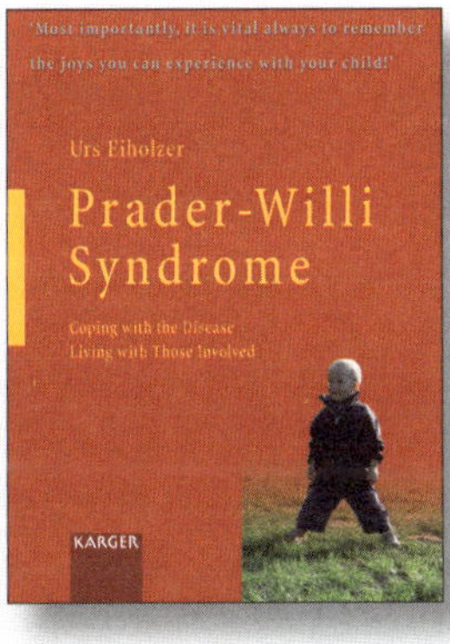

Stereoatlas of Ophthalmic Pathology

Anatomy and Pathology of the Peripheral Fundus (Fundus extremus)

A bequest of Basil Daicker, former Professor of Ophthalmic Pathology in Basel
Includes stereo glasses and a CD-ROM
Editors: *P. Meyer*, Basel; *K.U. Löffler*, Bonn

This stereoscopic atlas is based on the slide collection of the late Professor B. Daicker, Basel, who was renowned for his merits in ophthalmopathology the world over. From the compilation of on the whole 12,000 stereoscopic slides of interesting and rare eye diseases, the editors selected 246 pairs to make them available to interested health professionals. These images can be viewed through stereo glasses (provided with the atlas) to obtain a stereoscopic depth perception.

The atlas is divided into seven chapters. Each chapter starts with a short introduction, followed by the slides, which are complemented by a description of the disease and by additional images which can be found on the also included CD-ROM.

The technical efforts made together with the carefully selected content make this atlas unique and indispensable. The stereoscopical visualization enhances the comprehension of eye disorders and makes it a much-welcomed tool for the differential diagnostic process.

Stereoatlas of Ophthalmic Pathology
Anatomy and Pathology of the Peripheral Fundus (Fundus extremus)
A bequest of Basil Daicker, former Professor of Ophthalmic Pathology in Basel. Includes stereo glasses and a CD-ROM
Editors: Meyer P. (Basel), Löffler K.U. (Bonn)
XIV + 162 p., 246 fig., 240 in color, 1 tab., spiral binding, 2006
ISBN: 978-3-8055-7840-0
e-ISBN: 978-3-318-01160-9

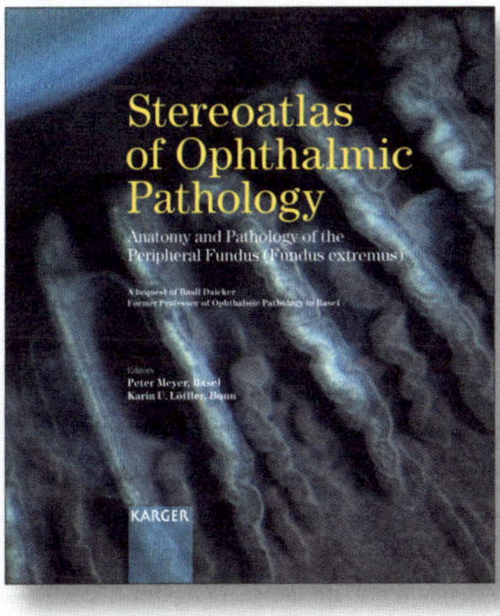

The Anatomy of the Human Embryo

A Scanning Electron-Microscopic Atlas

Author: *G. Steding*, Göttingen

The present anatomical atlas concentrates on the early weeks of prenatal development of the human embryo. It comprises more than 800 scanning electron-microscopic pictures of specimens of exclusively human embryos. The three-dimensional appearing illustrations show the development of the external form of the face, neck, trunk and limbs. Besides, the brain and the viscera of the head, neck, thorax, abdomen and pelvis – all dissected into layers – are represented in their position and spatial form.

The juxtaposition of pictures of temporally close developmental stages reveals the changes in the form of the organs. Photographs of the same organic system are usually shown at the same magnification and clearly demonstrate the growth process. Simple outline drawings provided with the principal nomenclature facilitate the orientation within the specimens. A brief introduction to each chapter explains the most significant developmental steps depicted.

This atlas is of great interest not only to anatomists, embryologists, histologists and developmental biologists, but also to biologists, biochemists and geneticists. Moreover, it serves as a valuable reference book for clinicians such as gynecologists, obstetricians, pediatric surgeons and pediatric cardiologists.

The Anatomy of the Human Embryo
A Scanning Electron-Microscopic Atlas
Author: Steding G. (Göttingen)
XIV + 516 p., 818 fig., hard cover, 2009
ISBN: 978-3-8055-8361-9

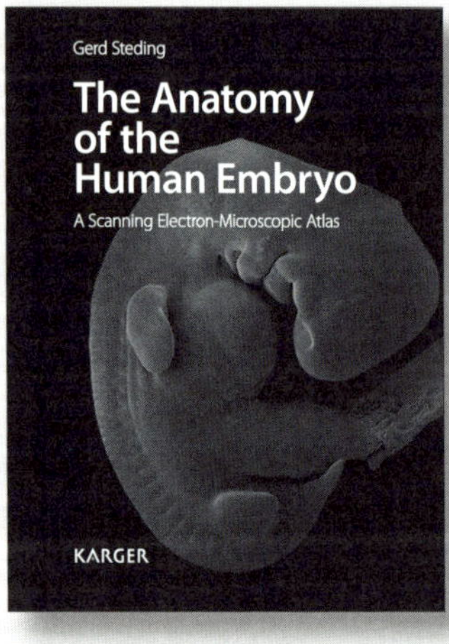

BMA

BMA Book Award
'highly commended' in
Basic and Clinical Sciences

The Road to Good Nutrition

Editors: *M. Eggersdorfer*, Kaiseraugst;
K. Kraemer, Basel; *M. Ruel*, Washington;
M. Van Ameringen, Geneva; *H.K. Biesalski*, Stuttgart;
M. Bloem, Roma; *J. Chen*, Changping District;
A. Lateef, Washington, D.C.; *V. Mannar*, Ottawa, Ont.

This is a work of advocacy, whose prime objective is to inform people about the relationship between nutrition security and public health. It draws on the thinking and experience of a selected number of experts in the field of nutrition and public health. Collating up-to-the-minute information in a clear and accessible way, the book forms a 'one-stop information source', and paves the way for further, science-led publications in this field. *The Road to Good Nutrition* puts the topic of nutrition security on the agenda of policy-makers, academics, private sector organizations and civil society, as well as of organizations dedicated to the nutrition space. It is also of interest to the educated lay reader who is generally well informed in matters of health, nutrition and sustainability.

The Road to Good Nutrition
Editors: Eggersdorfer M. (Kaiseraugst), Kraemer K. (Basel),
Ruel M. Washington, D.C.), Van Ameringen M. (Geneva),
Biesalski H.K. (Stuttgart), Bloem M. (Roma),
Chen J. (Changping District), Lateef A. (Washington, D.C.),
Mannar V. (Ottawa, Ont.),
XII + 210 p., 183 fig., 183 in color, 14 tab., soft cover, 2013
ISBN: 978-3-318-02549-1
e-ISBN: 978-3-318-02550-7

BMA

BMA Book Award
First Prize in Health and
Social Care

Trauma und Kunst/ Trauma and Art

Sexueller Missbrauch und Depression Sexual Abuse and Depression

Authors: *Rut*, Liestal; *G. Benedetti*, Riehen; *G. Waser*, Basel

Sexual abuse and the thus incurred enforced secrecy turn a victim's values and sense of order upside down. This severe psychological trauma may lead to defensive reactions such as a split personality, disturbed maturation process and depression.

This extraordinary book documents the therapeutic journey taken by Rut, a childhood victim of sexual abuse, from her suppressed childhood pain to her artistic expression of the trauma. Her pictorial work created during her psychotherapy enables her in her speechlessness and depression to recapture the story of her childhood, and, in dialogues with her psychotherapist, to integrate it into her everyday life. Rut's pictures and texts give direct evidence of the trauma of sexual abuse; they make it perceptible, arouse empathy, demonstrate its psychological consequences, and at the same time show a way to work through the trauma. In the second part of the book the two co-authors, both psychotherapists, discuss the creative process and pictorial work in an arttherapeutical and scientific setting, thus establishing a theoretical background. Unique and impressive, this book delineates part of Rut's psychological healing process, and is of great interest to professionals and students in psychology, psychotherapy, psychiatry and medicine, as well as art therapy, education, social work and spiritual counseling.

Trauma und Kunst/ Trauma and Art
Sexueller Missbrauch und Depression Sexual Abuse and Depression
Authors: Rut (Liestal), Benedetti G. (Riehen), Waser G. (Basel)
VIII + 136 p., 49 fig., 48 in color, hard cover, 2004
ISBN: 978-3-8055-7556-0

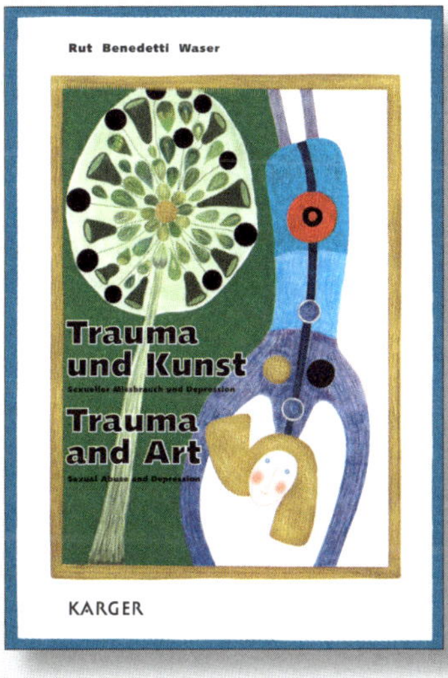

BMA
BMA Book Award 'highly commended' in Mental Health

Using and Understanding Medical Statistics

5th, revised and extended edition

Authors: *D.E. Matthews,* Waterloo, Ont.; *V.T. Farewell*, Cambridge

The fifth revised edition of this highly successful book presents the most extensive enhancement since *Using and Understanding Medical Statistics* was first published 30 years ago. Without question, the single greatest change has been the inclusion of source code, together with selected output, for the award-winning, open-source, statistical package known as R. This innovation has enabled the authors to de-emphasize formulae and calculations, and let software do all of the 'heavy lifting'.

This edition also introduces readers to several graphical statistical tools, such as Q-Q plots to check normality, residual plots for multiple regression models, funnel plots to detect publication bias in a meta-analysis and Bland-Altman plots for assessing agreement in clinical measurements. New examples that better serve the expository goals have been added to a half-dozen chapters. In addition, there are new sections describing exact confidence bands for the Kaplan-Meier estimator, as well as negative binomial and zero-inflated Poisson regression models for over-dispersed count data.

The end result is not only an excellent introduction to medical statistics, but also an invaluable reference for every discerning reader of medical research literature.

Using and Understanding Medical Statistics
Authors: Matthews D.E. (Waterloo, Ont.), Farewell V.T. (Cambridge)
5th, revised and extended edition
approx. XXIV + 332 p., 48 fig., 103 tab., hard cover + online supplement material, 2015
ISBN: 978-3-318-05458-3
e-ISBN: 978-3-318-05459-0

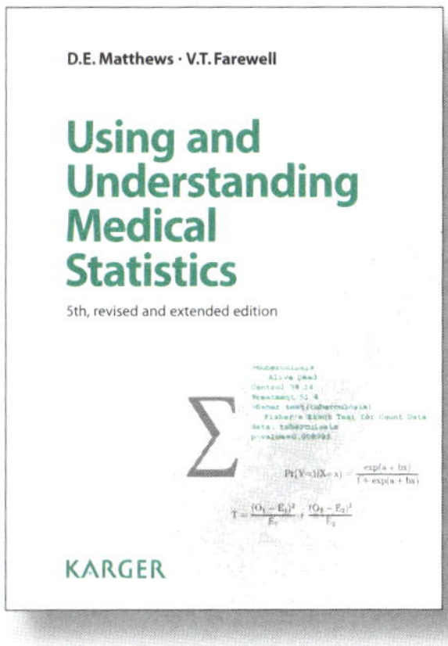

Sources of the Illustrations

14, 41, 43	Collection of Portraits, University Library, Basel
15	Brockhaus Enzyklopädie, ed. 17. Wiesbaden, 1971, vol. 13
20	Staehr, Ch: Spurensuche. Ein Wissenschaftsverlag im Spiegel seiner Zeitschriften, Stuttgart, Thieme, 1986
27	Album von Berlin: 2 grosse Panoramen und 51 Ansichten nach Momentaufnahmen in Photographiedruck. Berlin, Globus Verlag
32, 33, 101	Photographs by Gerri Schmid, Basel
35	Verlag Friedrich Reinhardt AG, Basel
40	Bildersammlung 3,362, Staatsarchiv Basel
48	The Roche Historical Collection and Archive, Basel
51	Ringier Dokumentationszentrum, Zürich
57	Lüttge U, Kluge M, Bauer G: Botanik. Ein grundlegendes Lehrbuch. Weinheim, VCH Verlagsgesellschaft, 1988
80	Photograph by Wolfram Gerlich, Marburg
94	Musée du Louvre, Paris

All other illustrations were taken from Karger publications and archives.

Impressum

Library of Congress Cataloging-in-Publication Data
Schmeck, Harold M., Jr. (Harold Marshall), 1923-2013, author.
Karger, connecting the world of biomedical science / Harold M. Schmeck, Jr., Mechthild Hempe.
p. ; cm.
Connecting the world of biomedical science
Festschrift in honor of Karger's 125th anniversary.
Expanded version of: Karger, turning medical progress into print / Harold M. Schmeck, Jr. c1990.
Includes bibliographical references.
ISBN 978-3-318-03033-4 (hard cover : alk. paper) -- ISBN 978-3-318-03034-1 (electronic version)
I. Hempe, Mechthild, author. II. Schmeck, Harold M., Jr. (Harold Marshall), 1923-2013. Karger, turning medical progress into print. Expanded version of (work): III. S. Karger (Firm), honouree. IV. Title. V. Title: Connecting the world of biomedical science.
[DNLM: 1. S. Karger (Firm) 2. Medicine--Festschrift. 3. Publishing--history--Festschrift. 4. History, 20th Century--Festschrift. 5. History, 21st Century--Festschrift. 6. Science--history--Festschrift. WZ 345]
Z430.S16
070.509494'32--dc23
2015021008

Bibliographic information published by the Deutsche Nationalbibliothek
The Deutsche Nationalbibliothek lists this publication in the Deutsche Nationalbibliografie; detailed bibliographic data are available on the Internet at http://dnb.dnb.de

Graphic Design and Typesetting: Gerri Schmid
Production Editor: Stefan Sessler
Copy Editor: Yannik Aeby, Andrea Shaw
Proof Reader: Franziska Selz, Herbert Wäckerlin
Image Processing: Laszlo Madarasi
Graphic Production Manager: Patrik Brunner

Printed in Switzerland on acid-free and non-aging paper (ISO 9706)
by Werner Druck & Medien AG, Basel
ISBN 978-3-318-03033-4
e-ISBN 978-3-318-03034-1